THERMOPLASTIC – COMPOSITE MATERIALS

Edited by **Adel Zaki El-Sonbati**

Thermoplastic – Composite Materials
Edited by Adel Zaki El-Sonbati

CBS, Edition **2017**

Published by InTech

Janeza Trdine 9, 51000 Rijeka, Croatia

Publishing Process Manager Oliver Kurelic

Technical Editor Teodora Smiljanic

Cover Designer InTech Design Team

Additional hard copies can be obtained from orders@intechopen.com

Thermoplastic – Composite Materials, Edited by Adel Zaki El-Sonbati
p. cm.
ISBN 978-953-51-0310-3

INTECH

open science | open minds

Contents

Preface VII

Chapter 1 **Novel Thermoplastic Polyimide Composite Materials 1**
Haixia Yang, Jingang Liu, Mian Ji and Shiyong Yang

Chapter 2 **Thermoplastic Polyurethanes-Fumed Silica Composites: Influence of NCO/OH in the Study of Thermal and Rheological Properties and Morphological Characteristics 11**
José Vega-Baudrit, Sergio Madrigal Carballo and José Miguel Martín Martínez

Chapter 3 **Manufacture of Different Types of Thermoplastic 25**
Lavinia Ardelean, Cristina Bortun, Angela Podariu and Laura Rusu

Chapter 4 **High Performance Thermoplastic/Thermosetting Composites Microstructure and Processing Design Based on Phase Separation 49**
Yuanze Xu and Xiujuan Zhang

Chapter 5 **Processing of Carbon Fiber/PEI Composites Based on Aqueous Polymeric Suspension of Polyimide 91**
Liliana Burakowski Nohara, Geraldo Maurício Cândido, Evandro Luís Nohara and Mirabel Cerqueira Rezende

Chapter 6 **Thermoplastic Nanocomposites and Their Processing Techniques 113**
Sajjad Haider, Yasin Khan, Waheed A. Almasry and Adnan Haider

Chapter 7 **Crystallization and Thermal Properties of Biofiber-Polypropylene Composites 131**
M. Soleimani, L. Tabil, S. Panigrahi and I. Oguocha

Preface

Composite materials are being more frequently used in a wide range and variety of structures, such as automotive and aerospace components. Composite materials often demand a unique combination of properties, including high thermal and oxidative stability, toughness, solvent resistance and low dielectric constant. This book is comprised of seven excellent chapters, written for all specialized scientists and engineers dealing with characterization, thermal, mechanical and technical properties, rheological, morphological and microstructure properties, and processing design of composite materials.

Chapter 1 reports novel thermoplastic polyimides (TPI) resins reinforced with carbon fiber (CF), glass fiber (GF), and modified by adding of solid lubricates such as graphite (Cr), poly(tetrafluoroethylene) (PTFE) or molybdenum disulfide (MoS2) to give TBI molding particulates, which could be injection-molded at elevated temperature to give the TPI composite materials. The thermal and mechanical properties of the pure TPI resin, the molding particulates and the molded composites were systematically characterized.

Thermoplastic polyurethanes (TPU`s) are a multipurpose group of phase segmented polymers that have good mechanical and elastic properties and hardness. Usually, TPU`s exhibits a two-phase microstructure. Fumed nanosilicas are added to increase the thermal, rheological and mechanical properties of TPU`s. Chapter 2 studies the effect of incorporating hydrophilic fumed nanosilica in the formulation of polyurethane adhesive with different NCO/OH to improve its thermal, rheological and adhesive properties.

Chapter 3 involves manufacturing techniques developed for composite structural insulated panels (CSIPs) on the construction site. Detailed description about the manufacturing CSIPs is included in this Chapter. The Chapter also covers manufacturing of the traditional structural insulated panels (SIPs) in the panelized construction.

Cure induced phase separation (SIPS) is an important part of reaction-induced phase separation (RIPS) due to its innovative applications to composite processing, or, more generally, to the innovations of multi-phase polymers. Chapter 4, provides a concise summary of up-to-data original contribution relevant literatures in this field emphasizing the breakthrough in approaches to understanding and controling the CIPS during the process.

Thermoplastics have some distinct advantages over thermoset composites, such as high ductility and toughness, facility of processing and recycling potential. The purpose of Chapter 5 is to compare two methods of processing thermoplastic composite hot compression molding and aqueous suspension prepregging, showing that the latter method uses the insertion of a polyimide interface in the composite.

Chapter 6 gives the reader a complete understanding of the thermoplastic nanocomposites and their processing techniques, polymer interaction, their resulting properties and proposed application.

In Chapter 7, differential scanning calorimetry (DSC) and heat flowmeter method are used to determine the non-isothermal crystallization behavior and thermal conductivity of polypropylene (PP) and its composites, reinforced or filled with different weight of fractions of biofibers.

The editor of this book would like to express his gratitude to Prof. M.A. Diab and Prof. A.A. El-Bindary, Chemistry Department, Faculty of Science (Demiatta), Mansoura University, Demiatta, Egypt, for their useful advice in the process of preparation of the book.

Prof. A.Z. El-Sonbati
Chemistry Department, Faculty of Science, Demiatta,
Mansoura University, Demiatta,
Egypt

1

Novel Thermoplastic Polyimide Composite Materials

Haixia Yang*, Jingang Liu, Mian Ji and Shiyong Yang*
Laboratory of Advanced Polymer Materials, Institute of Chemistry Chinese Academy of Sciences, Beijing, China

1. Introduction

Novel thermoplastic polyimide (TPI) offer several potential advantages over thermoset polyimides. First, TPI has an indefinite shelf life, low moisture absorption, excellent thermal stability and chemical resistance, high toughness and damage tolerance, short and simple processing cycles and potential for significant reductions in manufacturing costs. Second, they have the ability to be re-melt and re-processed, thus the damaged structures can be repaired by the applying heat and pressure. Thirdly, TPI offer advantages in the environmental concerns. Usually, TPI has very low toxicity since it is the completely imidized polymer, does not contain any reactive chemicals. Due to the re-melt possibility by heating and re-dissolvability in solvents, TPI could be recycled or combined with other recycled materials in the market to make new products [1-3].

For injection or extrusion moldings, conventional polyimides do not have enough flow properties, therefore, only limited fabrication processes such as compression, transfer or sintering molding could be applied. Significant efforts have been devoted to develop melt processable polyimides. Most of the efforts have been focused on exploiting of the correlation between chemical structures and polymer properties, such as Tg and melting ability etc. to improve TPI's melt processability, thus resulted in some commercial TPI materials such as amorphous LARC-TPI resin (Tg ~ 250 °C), ULTEM ® resin (Tg ~ 217 °C) and semi-crystalline Aurum® resins (Tg ~ 250 °C and Tm ~ 380 °C). Bell and St. Chair investigated the effect of diamine structure on the thermal properties of polyimides and such studies led to the invention of LARC-TPI [4-7]. The characteristic structure of LARC-TPI, meta-substituted diamine and the flexible linkage between the benzene rings enhanced the thermoplasticity. To make the polyimides more processable, there have been reports of several modification of polymer structure [8, 9]. In recent years, a new developed TPI was commercialized as EXTEM® XH and UH resins by SABIC Innovative Plastics. With its high temperature capability (Tg ~ 267 °C and 311°C) and high melt flow ability, EXTEM resin differentiates its position within TPIs as well as other high performance polymers [10, 11].

In this paper, a series of novel TPI composites have been prepared and their thermal, rheological and mechanical properties were characterized. The TPI resins have excellent

* Corresponding Authors

melt flow capability, which can be fiber-reinforced or filler-modified to give high quality a series of TPI composites. Moreover, very thin-walled complex parts can be injection molded at elevated temperatures.

2. Experimental

2.1 Materials

1,4-Bis(4-amino-2-trifluoromethylphenoxy)benzene (6FAPB) was synthesized in this laboratory according to the reported methods [12, 13]. 4,4-Oxydiphthalic anhydride (ODPA) was purchased from Shanghai Research Institute of Synthetic Resins, China and dried in a vacuum oven at 180 °C for 12 h prior to use. Phthalic anhydride (PA) was commercially purchased and sublimed prior to use. *N*-methyl-2-pyrrolidinone (NMP) was purified by vacuum distillation over P_2O_5 prior to use. Toluene (Beijing Beihua Fine Chemicals Co., China) was used as received without further purification. Carbon fiber (T800) was purchased from Toray and used as received. Glass fiber was purchased from Beijing Xingwang Glass Fiber Ltd. Corp. and used as received. Graphite, poly(tetrafluoroethylene) (PTFE) and molybdenum disulfide (MoS_2) were afforded by Beijing POME Corp. and used as received.

2.2 Measurements

Differential scanning calorimetry (DSC) was performed on a TA Q100 thermal analysis system in nitrogen atmosphere at a flow rate of 50 cm^3/min and the scanning range was from 50 to 350 °C. The glass transition temperature (Tg) was determined by the inflection point of the heat flow versus temperature curve. Complex viscosity (η^*) were measured on a TA AR2000 rheometer. A TPI resin desk with 25 mm in diameter and ~1.5 mm in thickness was prepared by press-molding the resin powder at room temperature, which was then loaded in the rheometer fixture equipped with 25 mm diameter parallel plates. Measurements were performed using the flow mode with a constant stress (104 Pa) and about 5 N normal forces. In the temperature ramp procedure, an initial temperature of 200 °C was set and then the parallel plates with testing sample were equilibrated at this temperature for 10 min. The complex viscosity (η^*) as a function of the scanning temperature (T) were measured by scanning the temperature from 200 °C to 400 °C at a rate of 3 °C /min.

Thermal gravimetric analysis (TGA) and the coefficients of thermal expansion (CTE) were performed on a Perkin-Elmer 7 Series thermal analysis system at a heating rate of 20 °C /min in nitrogen atmosphere at a flow rate of 30 cm^3/min. Dynamic mechanical analysis (DMA) was performed on a Perkin-Elmer 7 Series thermal analysis system, and the scanning temperature range was from 50 °C to 320 °C at a heating rate of 5 °C/min and at a frequency of 1 Hz. A three-point bending mode was employed and the specimen size was 15.0 × 3.0 × 1.2 mm^3. The storage modulus (G'), loss modulus (G'') and tangent of loss angle (tanδ) were obtained as the function of scanning temperature. Melt flow index (MFI) were measured in according with GB/T3680-2000 at elevated temperature.

The mechanical properties were measured on an Instron-5567 universal tester at different temperature. The tensile strength, modulus, and elongation at break were measured in according with GB/T16421-1996 at a strain rate of 2 mm/min. The flexural strength and modulus were measured in according with GB/T5270-1996 at a strain rate of 2 mm/min.

The compressive strength and modulus were measured in according with GB/T2569-1997 at a strain rate of 2 mm/min. Izod impact (unnotched) were measured in according with GB/T16420-1996.

2.3 Preparation of the TPI resins

TPI with controlled molecular weights was prepared by the reaction of ODPA with 6FAPB in the presence of PA as endcapping agent in NMP at elevated temperatures (scheme 1) [14-17]. 6FAPB (2161.65 g, 5.047 mol) and NMP (21 L) were placed into a 50 L agitated reactor equipped with a mechanical stirrer, a thermometer, nitrogen inlet/outlet and a condenser. The mixture was stirred at ambient temperature for ~1 h until the aromatic diamine was completely dissolved to give a homogeneous solution. ODPA (1508.00 g, 4.859 mol) and PA (41.62 g, 0.281 mol) were then added. An additional 0.6 L of NMP was used to rinse all of the anhydrides, resulting in a mixture with 15% solid content (w/w). After the mixture was stirred in nitrogen at 75 °C for 4 h, 2 L of toluene and a few drops of isoquinoline as a catalyst were added. The obtained solution was gradually heated to 180 °C and held for 10 h with stirring. The water evolved during the thermal imidization was removed simultaneously by azeotropic distillation. After the thermal imidization reaction was completed, the reaction solution was cooled down to room temperature and then poured into excess of ethanol with vigorous stirring to precipitate the polyimide resin. The solid resin was then isolated by filtration, thoroughly washed with warm ethanol and dried at 100 °C overnight to remove most of the ethanol. The polyimide resin was fully dried at 205 °C in a vacuum oven for ~24 h to give 3420 g (97%) of TPI.

2.4 Preparation of TPI composites

The pure TPI resin powder was dried at 205°C for 6 h in a vacuum dryer, and then extruded at elevated temperature with carbon fiber, glass fiber, graphite, poly(tetrafluoroethylene) (PTFE) or molybdenum disulfide (MoS_2) to give composite molding particulates, which were abbreviated as CF-TPI, GF-TPI, Gr-TPI, PTFE-TPI and MoS_2-TPI, respectively, as shown in Table 1. The composite molding particulates could be injection-molded at elevated temperature to give the TPI composites. The test composites samples were injected on a standard 120-ton injection molding machine equipped with a general purpose screw in accordance to the guidelines presented in Table 2.

3. Results and discussion

3.1 Preparation of TPI resins and TPI composites

The TPI resin with designed polymer backbones and controlled molecular weights were prepared by a one-step thermal polycondensation procedure as shown in Scheme 1. The offset of the aromatic dianhydride (ODPA) to the aromatic diamine (6FAPB) and endcapping agent (PA) was used to control the polymer molecular weights. The water evolved during the thermal imidization was removed simultaneously from the reaction system by azeotropic distillation. The TPI resin showed several characteristic absorption in FT-IR spectra, including the absorptions at 1780 and 1720 cm^{-1} attributed to the asymmetrical and symmetrical stretching vibrations of the imide groups, the band at 1380 cm^{-1} assigned as the C–N stretching vibration, and the absorptions at 1100 and 725 cm^{-1} due to the imide ring deformation, etc.

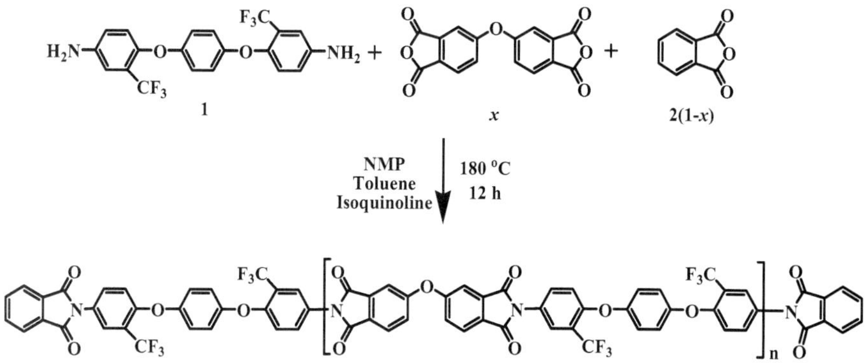

Scheme 1. Synthesis of thermoplastic polyimide resin

The chemical compositions of the TPI composites were listed in Table 1. Firstly, the pure TPI resin powder was dried at 205 °C for 6h in a vacuum dryer to completely remove the moisture in the resin, which was then extruded at elevated temperatures with carbon fiber, glass fiber, MoS_2 or PTFE to afford the TPI molding particulates. The processing parameters for the injection of the TPI molding particulates were shown in Table 2, in which the melt temperature was settled at 350 – 370 °C and the molding temperature at 150 – 160 °C.

	TPI	Carbon Fiber	Glass Fiber	Graphite	MoS_2	PTFE
TPI	100	-	-	-	-	-
CF-TPI-10	90	10	-	-	-	-
CF-TPI-20	80	20	-	-	-	-
CF-TPI-30	70	30	-	-	-	-
GF-TPI-15	85	-	15	-	-	-
GF-TPI-30	70	-	30	-	-	-
GF-TPI-45	55	-	45	-	-	-
Gr- TPI-15	85	-	-	15	-	-
Gr- TPI-40	60	-	-	40	-	-
MoS_2-TPI-15	85	-	-	-	15	-
MoS_2-TPI-30	70	-	-	-	30	-
PTFE-TPI-20	80	-	-	-	-	20

Table 1. Chemical compositions of the TPI composites

Processing Conditions	Unit	Parameter
Drying Conditions	°C/hr	160/4~6
Moisture Content (max.)	%	0.02
Melt Temperature	°C	350~370
Mold Temperature	°C	150~160
Back Pressure	MPa	0.3~0.5
Screw Speed	RPM	40~80

Table 2. Injection molding conditions of the molding particulates

3.2 Rheological properties of the TPI molding particulates

Dynamical rheology was employed to investigate the melt properties of the molding particulates. Figure 1 compares the melt viscosities of carbon fiber-filled TPI resins with different loadings at different temperature and the dates are summarized in Table 3. It can be seen that the molten viscosities of molding particulates decreased gradually with increasing the temperature scanned at 200-400°C, primarily attributed to the melting of TPI resin in the molding particulates. The minimum melt viscosities of the molding particulates increased with increasing of the carbon fiber loadings. For instance, the minimum melt viscosity of CF-TPI-10 was 1.8×10^3 Pa·s at 400 °C, lower than that of CF-TPI-30 (6.7×10^3 Pa·s at 400 °C). Meanwhile, the melt viscosity at the processing temperature (360 °C) was increased from 4.7×10^3 Pa·s for CF-TPI-10 to 9.4×10^3 Pa·s for CF-TPI-30, indicating that the addition of carbon fiber in TPI resins increased the melt viscosities of the molding particulate, thus lowering their melt processabilities. The melt viscosities of other TPI composites filled with glass fiber, graphite, molybdenum disulfide (MoS_2) and poly(tetrafluoroethylene) (PTFE) are also shown in Table 3. It can been seen that the molding particulates (Gr-TPI, MoS_2-TPI and PTFE-TPI) showed good melt processabilities with complex melt viscosities of 3.6×10^3 Pa.s (Gr-TPI-15) at 360 °C, 4.1×10^3 Pa.S (MoS_2-TPI-15) at 360 °C, and 4.5×10^3 Pa·s (PTFE-TPI-20) at 360 °C, respectively. It should be noted that glass fiber-filled TPI molding particulates showed much higher complex melt viscosities than the pure TPI resin. For instance, GF-TPI-30 has a complex melt viscosity of 4.5×10^4 Pa·s at 360 °C, 4.7 times higher than CF-TPI-30 (9.4×10^3 Pa·s at 360 °C).

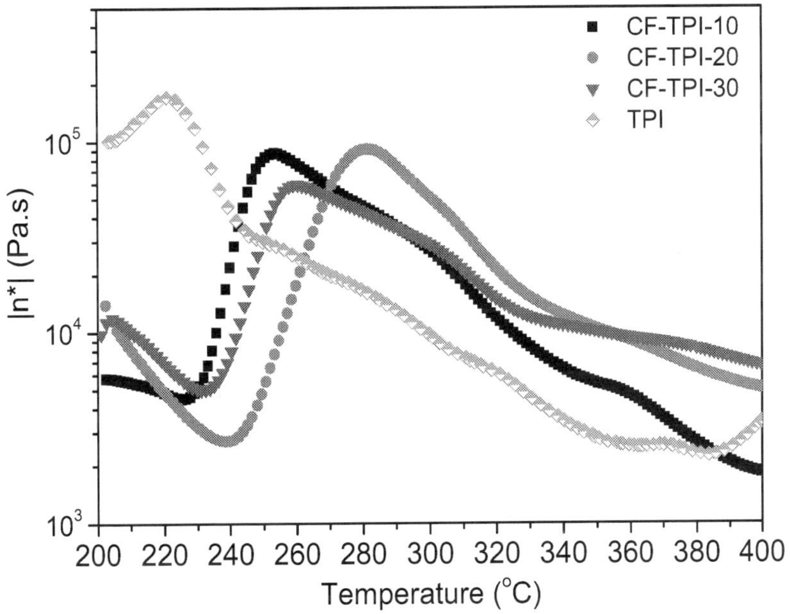

Fig. 1. Dynamic rheological behaviors of the carbon fiber-filled TPI molding particulates

	Minimum melt viscosities (Pa.s)	Complex melt viscosities (Pa.s) at 360 °C
TPI	2.5×10^3 at 360 °C	2.3×10^3
CF-TPI-10	1.8×10^3 at 400 °C	4.7×10^3
CF-TPI-20	5.1×10^3 at 400 °C	9.1×10^3
CF-TPI-30	6.7×10^3 at 400 °C	9.4×10^3
GF-TPI-15	8.7×10^3 at 340 °C	3.4×10^4
GF-TPI-30	1.5×10^4 at 300 °C	4.5×10^4
Gr-TPI-15	2.4×10^3 at 280 °C	3.6×10^3
Gr-TPI-40	4.4×10^3 at 330 °C	6.3×10^3
MoS_2-TPI-15	3.0×10^3 at 400 °C	4.1×10^3
MoS_2-TPI-30	4.0×10^3 at 290 °C	4.9×10^3
PTFE-TPI-20	3.4×10^3 at 280 °C	4.5×10^3

Table 3. Melt Viscosities of TPI and TPI-c molding particulates at Elevated Temperature

Melt flow index of the molding particulates was shown in Table 4. The pure TPI resin showed highest melt flow index of 10.1 g/10 min at 360 °C under a pressure of 10 kg, which could be employed to inject very thin-walled complex parts due to its excellent melt flow ability. Figure 2 shows a representative thin-walled complex part which has a wall thickness of 0.2mm. In general, the TPI resins filled with different fillers such as graphite, molybdenum disulfide and poly(tetrafluoroethylene) etc. all showed good melt flow properties with melt flow index of > 2.0 g/10min. In comparison, the carbon fiber-filled TPI molding particulates showed better melt flow properties than the Glass fiber-filled ones. For instance, CF-TPI-30 has a melt flow index of 2.4 g/min at 360 °C under 10 kg, compared with GF-TPI-30 (0.5 g/min at 360 °C under 21.6 kg. Meanwhile, the filler loadings also have obvious effect on lowering the melt flow index. For instance, MoS_2-TPI-30 has a melt flow index of 0.8g/10 min at 360 °C under 10 kg, much lower than MoS_2-TPI-15(3.2 g/10 min at 360 °C under 10 kg).

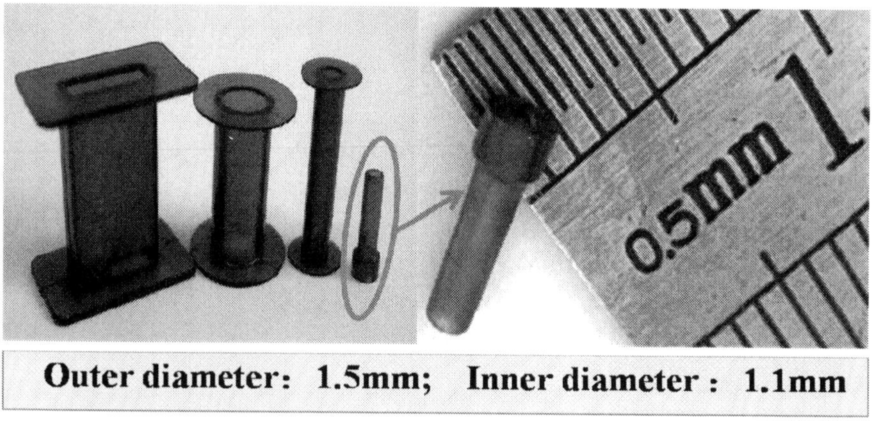

Fig. 2. The Injection molded TPI thin-walled parts

	Temperature/ °C	M/kg	t/s	g/10min
TPI	360	10	10	10.1
CF-TPI-10	360	10	30	3.8
CF-TPI-20	360	10	30	2.5
CF-TPI-30	360	10	30	2.4
GF-TPI-15	360	21.6	60	0.9
GF-TPI-30	360	21.6	60	0.5
Gr-TPI-15	360	10	10	8.9
Gr-TPI-40	360	10	30	6.0
MoS_2-TPI-15	360	10	30	3.2
MoS_2-TPI-30	360	10	60	0.8
PTFE-TPI-20	360	10	60	2.4

Table 4. Melt Flow Index at 360 °C of the Molding Particulates

3.3 Thermal properties of the TPI molded composites

Figure 3 compares DSC curves of the injection-molded TPI composites with different carbon fiber loadings. It can be seen that molded composites showed glass transition temperature (T_g) in the range of 215-216 °C, which was not obviously changed by the carbon fiber loadings. Figure 4 shows the DMA curves of a representative molded composite (CF-TPI-20), in which the peak temperature in the Tan δ curve was at 211 °C. The storage modulus curve did not turn down until the temperature was scanned up to 201 °C, demonstrating that the carbon fiber-filled TPI molded composites have outstanding thermo-mechanical properties. Figure 5 depicts the thermal stabilities of carbon fiber-filled TPI molded composites. The temperatures at 5% and 10% of original weight losses were measured at 550 °C and 580 °C, respectively. The initial decomposition temperatures were determined at 550 °C and the char yields at 750 °C was > 60%.

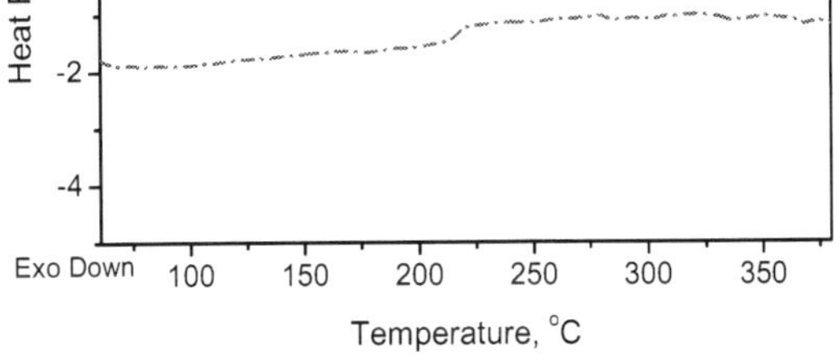

Fig. 3. DSC curves of carbon fiber-filled TPI molded composites

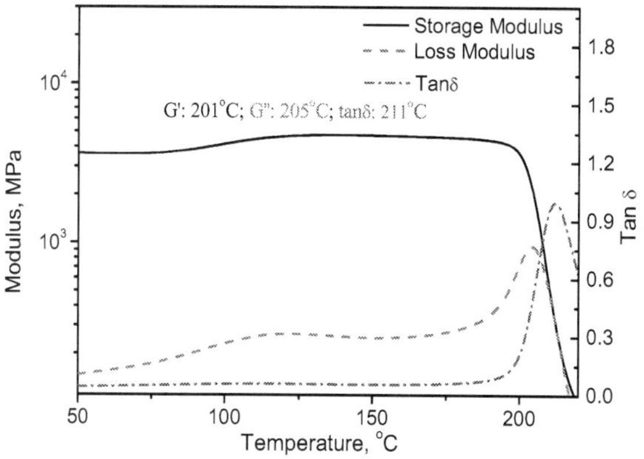

Fig. 4. DMA curves of CF-TPI-20 molded composite

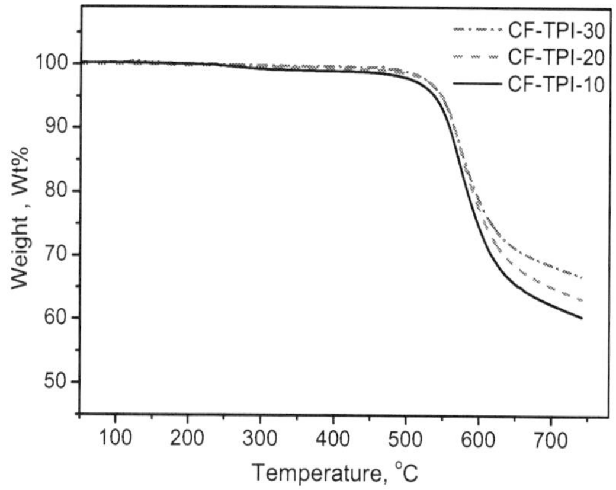

Fig. 5. TGA curves of carbon fiber-filled TPI molded composites

3.4 Mechanical properties of the molded composites

Table 5 compares the mechanical properties of the injection molded TPI composites. The pure TPI resin exhibited good combined mechanical properties with tensile strength of 100 MPa, tensile modulus of 5.6 GPa, elongation at break of 57.6%, flexural strength of 154 MPa, flexural modulus of 3.8 GPa, and izod impact (un-notched) of 156 kJ/m^2. The carbon fiber-filled TPI molded composites possess mechanical properties better than the pure TPI resin with tensile strength of 177-219 MPa, tensile modulus of 7.3-12.4 GPa, flexural strength of 241-327 MPa and Izod impact of 20.8-24.4 kJ/m^2, demonstrating that carbon fiber have significant reinforcing effect. The glass fiber-filled TPI molded composites also showed good mechanical strength, but lower modulus than the carbon fiber-filled ones. For instance, GF-TPI-30 has a modulus of 7.5 GPa, only 60% of CF-TPI-30 (12.4 GPa). The other TPI molded composites filled with graphite, molybdenum disulfide and poly(tetrafluoroethylene) all showed good combined mechanical properties, demonstrating that the addition of filler did not deteriorate the TPI mechanical properties.

	Tensile strength (MPa)	Tensile modulus (GPa)	Elongation at breakage (%)	Flexural strength (MPa)	Flexural modulus (GPa)	Compressive strength (MPa)	Compressive modulus (GPa)	Izod impact (unnotched) (KJ/m^2)
TPI	100	5.6	57.6	154	3.8	159	3.6	156
CF-TPI-10	177	7.3	3.40	241	8.8	168	5.4	20.8
CF-TPI-20	177	10.1	2.14	278	13.5	215	7.6	20.8
CF-TPI-30	219	12.4	2.34	327	17.8	205	7.2	24.4
GF-TPI-15	121	4.8	3.42	191	5.7	145	3.7	20.4
GF-TPI-30	137	7.5	2.20	211	9.2	182	5.3	23.1
GF-TPI-45	106	8.8	1.62	242	14.6	207	7.0	16.6
Gr-TPI-15	94	4.4	4.37	150	5.9	106	2.7	15.4
Gr-TPI-40	64	8.2	1.71	114	11.2	86	3.9	6.7
MoS_2-TPI-15	102	8.9	13.8	148	3.8	116	2.7	23.7
MoS_2-TPI-30	83	4.1	3.36	145	5.6	115	2.9	13.1
PTFE-TPI-20	98	3.0	6.64	156	4.3	99	2.8	26.9

Table 5. Mechanical properties of the TPI molded composites

4. Conclusions

Novel thermoplastic polyimide (TPI) resins with designed polymer backbones and controlled molecular weights have been synthesized by thermal polycondensation of aromatic dianhydrides and aromatic diamines in presence of endcapping agent. The TPI resins were reinforced with carbon fiber, glass fiber, or modified by adding of solid lubricants such as graphite, poly(tetrafluoroethylene) (PTFE) or molybdenum disulfide (MoS_2) to give TPI molding particulates, which could be injection-molded at elevated temperature to give the TPI molded composites. Thus, thin-walled molded parts could be fabricated. The TPI molding particulates showed excellent melt processibility to produce high quality TPI molded composites with excellent combination of thermal and mechanical properties.

6. Acknowledgements

The author would like to thank the Contract grant sponsor: National Natural Science Foundation of China, contract grant number: 50903087

5. References

[1] Arnt R. Offringa, Composites: Part A, 329-336, 1996.
[2] Lars A. Berglund, Handbook of Composites, Edited by S.T. Peters, 115-130, 1998.
[3] Anne K. St. Clair, Terry L. St. Clair, "A multi-Purpose Thermoplastic Polyimide", SAMPE Quarterly, October, 20-25, 1981.
[4] Akihiro Yamaguchi, Masahiro Ohta, SAMPE Journal, 28-32, January/February, 1987.
[5] T.L. St. Clair, D.J. Progar, Polymer Preprint, 10,538, 1975.
[6] B.V Fell, J. Polymer Science, Polymer Chemistry Edition, 14, 2275, 1976.

[7] A.K. St. Clair, T.L. St. Clair, SAMPE Quarterly, 13(1), 20, 1981.
[8] H.D. Burks, T.L. St. Clair, J. Applied Polymer Science, 30, 2401, 1985.
[9] S. Maudgal, T.L. St. Clair, Int. Adhesion and Adhesives, 4(2), 87, 1984.
[10] S. Montgomery, D. Lowery, and M. Donovan, SPE Antec Tech. Conf., 2007.
[11] Kapil C. Sheth, "Highest heat amorphous thermoplastic polyimide blends", SPE Antec Tech. Conf., 2009.
[12] Xie, K., Zhang, S.Y., Liu, J.G., He, M.H. and Yang, S.Y., J. Polym. Sci., Part A: Polym. Chem., 39: 2581-2590, 2001
[13] Xie, K., Liu, J.G., Zhou, H.W., Zhang, S.Y., He, M.H. and Yang, S.Y., Polymer, 42: 7267-7274, 2001.
[14] Hongyan Xu, Haixia Yang, Liming Tao, Lin Fan, Shiyong Yang, Journal of Applied Polymer Science, Vol. 117, 1173-1183, 2010.
[15] Hongyan Xu, Haixia Yang, Liming Tao, Jingang Liu, Lin Fan, Shiyong Yang, High Performance Polymer, Vol. 22, 581-597, 2010.
[16] Wang, K., Yang, S.Y., Fan, L., Zhan, M.S. and Liu, J.G., J. Polym. Sci., Part A: Polym. Chem., 44: 1997-2006, 2006
[17] Wang, K., Fan, L., Liu, J.G., Zhan, M.S. and Yang, S.Y., J. Appl. Polym. Sci., 107: 2126-2135, 2008

2

Thermoplastic Polyurethanes-Fumed Silica Composites: Influence of NCO/OH in the Study of Thermal and Rheological Properties and Morphological Characteristics

José Vega-Baudrit[1,2], Sergio Madrigal Carballo[2]
and José Miguel Martín Martínez[3]
[1]*Laboratorio Nacional de Nanotecnología LANOTEC-CeNAT,*
[2]*Laboratorio de Polímeros POLIUNA-UNA,*
[3]*Laboratorio de Adhesión y Adhesivos, Universidad de Alicante,*
[1,2]*Costa Rica*
[3]*España*

1. Introduction

Thermoplastic polyurethanes (TPU´s) are a multipurpose group of phase segmented polymers that have good mechanical and elastic properties and hardness. Usually, TPU´s exhibit a two-phase microstructure, which arises from the chemical incompatibility between the soft and the hard segments. The hard rigid segment segregates into a glassy or semicrystalline domain and the polyol soft segments form amorphous or rubbery matrices, in which the hard segments are dispersed (Oertel, 1993). Many factors influence in the separation of phases as the molecular weight, the segmental length, the crystallizability of the segment, the overall composition and the intra- and inter-segments interactions. Fumed nanosilicas are added to increase the thermal, rheological and mechanical properties of TPU´s (Maciá-Agulló et al., 1992; Jaúregui-Belogui et al., 1999; Jaúregui-Belogui et al., 1999; Torró-Palau et al., 2001, Péres-Limaña et al., 2001).

When hydrophilic fumed nanosilica is added, the degree phase separation increases due to the interaction hydrogen-bonded between silanol groups on the nanosilica surface and soft segments of the TPU. Therefore, the segmental incompatibility on the TPU is increased with the presence of the hydrophilic nanosilicas. Recent studies have demonstrated that the use of this kind of materials able to form hydrogen-bonds result in less direct interactions between phases, causing a higher phase separation. Furthermore, the interactions between silanol and carbonyl groups are weaker than those between NH and ester carbonyl groups, then silica addition increases the polyester chain mobility and, it allows to become more ordered in relation to the TPU without silica (Sánchez-Adsuar et al., 2000; Tien et al., 2001; Nunes et al., 2000; Nunes et al., 2001).

The aim of this paper is to study the effect of incorporating hydrophilic fumed nanosilica in the formulation of polyurethane adhesives with different NCO/OH to improve its thermal, rheological and adhesive properties. The hypothesis is that the degree of polyurethane phase segregation was affected by the presence of silica and the formation of hydrogen bonds. Therefore, there should be a variation of properties in polyurethanes in response to the presence of dispersed silica.

In recent papers are showed the results of evaluating these samples using thermal, rheological and mechanical analysis, and adhesion tests (Vega-Baudrit et al., 2006; Navarro-Bañón et al., 2005; Vega-Baudrit et al., 2008; Vega-Baudrit et al., 2009).

2. Materials and methods

Fumed silica (nanosilica HDK N20) was manufactured by Wacker-Chemie (Burghausen, Germany). The nominal primary particle size in all nanosilicas was 7 nm. According to Wacker-Chemie, the nominal specific surface area of all nanosilicas was 200m^2/g and 100% of silanol groups.

The TPU was prepared using the prepolymer method. The prepolymer was obtained by reacting the polyadipate of 1,4-butanediol (M_w = 2440 Daltons) with 4,4-diphenyl methane diisocyanate - MDI; using different isocyanate/macroglycol equivalent ratios (1,05; 1,15; 1,25). 1,4-butanediol was used as chain extender. High purity solid MDI was supplied by Aldrich (Cat. 25.643-9), a mixture of 98 wt% of the 4,4′-isomer and 2 wt% of the 2,4′-isomer. The NCO content of the prepolymer was determined by titration with dibutylamine (UNE-EN 1242 standard). The polyadipate of 1,4-butanediol (Hoopol F-530) was supplied by Hooker S.A. (Barcelona, Spain) and was heated for 4 hours at 70°C under reduced pressure (5 Torr) to remove the residual water. The 1,4-butanediol was supplied by Aldrich (Cat. B8, 480-7) and was dried using 4 Å molecular sieves. To avoid crosslinking reactions during polyurethane synthesis, the reaction temperature was kept below 65°C under a stirring speed of 80 rpm. The synthesis of the polyurethane was carried out in dry nitrogen atmosphere to avoid the presence of water in the reactor. The prepolymers containing unreacted isocyanate ends were completely reacted with the necessary stoichiometric amount of 1,4-butanediol. The reaction time was 2 hours.

TPU solutions were prepared by mixing 20 wt% solid polyurethane and 2 wt% nanosilica with 2-butanone in a Dispermix DL-A laboratory mixer, provided with a Cowles mechanical stirrer (diameter = 50 mm) and a water jacket to maintain the temperature at 25°C during the preparation of the adhesives. This preparation was carried out in two consecutive stages: *i)* the nanosilica was mixed for 15 min at 2500 rpm with 1/3 butanone volume required. *ii)* the TPU and 2/3 butanone volume were added to the previous solution, stirring the mixture for 2h at 2000 rpm. TPU solutions were kept in a hermetic container until use. A TPU solution without silica was also prepared as control. Most of the properties of the polyurethanes were measured using solid films, which were prepared by placing about 100 cm^3 of solution in a mould and allowing a slow evaporation of the solvent at room temperature during 2 days. The polyurethane films obtained were about 0.7 to 0.9 mm thick. The nomenclature of the polyurethane-nanosilica mixtures were PU105, PU115 and PU125 (according with NCO/OH, respectively).

2.1 Experimental techniques

Samples were characterized by FTIR with Attenuated Total Reflectance (ATR), Differential Scanning Calorimetry DSC, Dynamic Mechanical Thermal Analysis DMTA, Transmission Electronic Microscopy TEM and X-ray Diffraction XRD.

The IR spectra of the polyurethane films were obtained in the transmission mode using a Bruker Tensor 27 spectrophotometer. Under the experimental conditions used, the signal/noise ratio of the equipment was 0.04% transmittance (at 2000 cm^{-1}). The resolution was 4 cm^{-1} and 80 scans were recorded and averaged.

DSC experiments were carried out in a TA instrumentDSC Q100 V6.2. Aluminium pans containing 12–15 mg of sample were heated from -80°C to 80°C under nitrogen atmosphere. The heating rate was 10 °C/min. The first heating run was carried out to remove the thermal history of the samples. From the second heating run, the glass transition temperature (Tg), the melting temperature (Tm), the crystallization temperature (Tc), the melting enthalpy (ΔHm), and the crystallization enthalpy (ΔHc) of the TPUs were obtained. The crystallization rate was estimated by melting the polyurethane film at 100 °C, followed by a sudden decrease to 25 °C and the evolution of heat with time under isothermal conditions was monitored for 30 min at 25 °C until a crystallization peak appeared.

The viscoelastic properties of the polyurethanes were measured in a Rheometric Scientific DMTA Mk III instrument using the two-point bending mode (single cantilever). The experiments were carried out by heating the sample from -80 °C to 100 °C, using a heating rate of 5 °C/min, a frequency of 1Hz and a strain of 64 mm peak–peak.

A JEOL JEM-2010 instrument was used to analyze the morphology of the nanosilicas; an acceleration voltage of100 kV was used. The nanosilicas were placed directly into the grid specially design for TEM analysis.

The polyurethane crystallinity was determined using Seifert model JSO-DEBYEFLEX 2002 equipment. This equipment was provided with a copper cathode and a nickel filter, and the monochromatic radiation of copper (Ka) was used as the X-ray source (λ=1,54Å). A range of diffraction angles (2θ) from 5° to 90° were used in the experiments.

3. Results and discussion

Recent studies (Nunes et al., 2000; Nunes et al., 2001; Vega-Baudrit et al., 2006; Navarro-Bañón et al., 2005; Vega-Baudrit et al., 2008; Vega-Baudrit et al., 2009) have demonstrated that the addition of fillers as silica able to form hydrogen-bonds result in less direct interactions between phases, causing a higher degree of phase separation in the polyurethane. On the other hand, the interactions between the silanol groups and the carbonyl groups in the polyurethane are weaker than those between the N-H and ester carbonyl groups, and therefore the silica addition increases the polyester chain mobility in the polyurethane allowing the creation of more ordered phases with respect to the polyurethane without silica.

3.1 Characterization of polyurethanes with different NCO/OH

Synthesized thermoplastic polyurethanes (TPU) with different NCO/OH were characterized by IR-FTIR spectroscopy (Figure 1). No significant differences among the TPU, except for a higher intensity of the bands between 900 and 1300 cm^{-1}, which increases with increasing the NCO/OH. Similarly, there is no characteristic band of the isocyanate groups (-NCO) close to 2250 cm^{-1}, indicating that the reaction was complete.

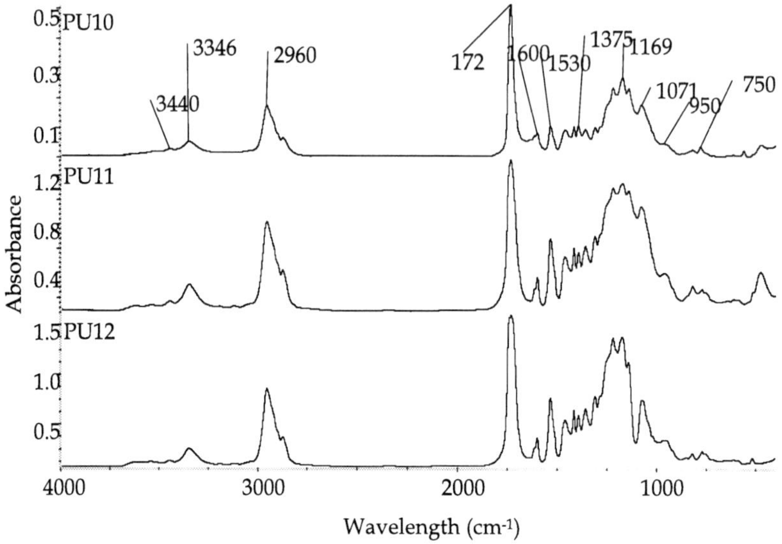

Fig. 1. FTIR of synthesized thermoplastic polyurethanes (TPU) with different NCO/OH.

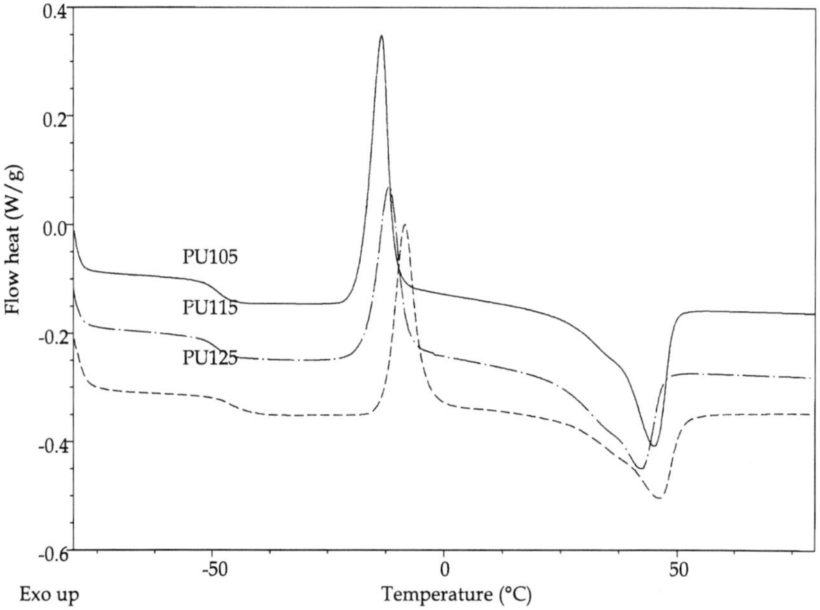

Fig. 2. DSC thermograms of synthesized thermoplastic polyurethanes (TPU) with different NCO/OH

In DSC thermograms (Figure 2), first glass transition temperature (T_{g1}) is close to -48 °C and is associated with soft segments of TPU. From -8 to -13 °C is showed the crystallization of soft segments, with an enthalpy of crystallization located between -20 and -26 J/g. Also, close to 46 °C, it shows the melting temperature of soft segments, with a melting enthalpy of approximately 26 J/g. Finally, a second DSC thermogram carried out up to 300 °C shows a second glass transition temperature (T_{g2}) close to 250 °C, which corresponds to the hard segments. Figure 3 shows parallel plate rheology - storage modulus (G') and loss (G'') as a function of temperature - of sample PU105.

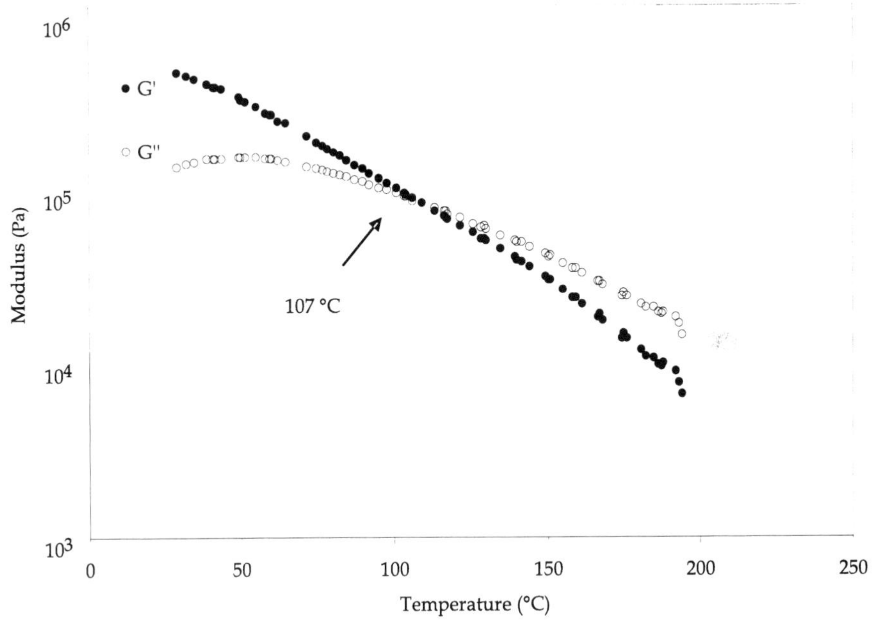

Fig. 3. Storage modulus (G') and loss (G'') of sample PU105.

Figure 4 shows the storage modulus G' for samples with different NCO/OH. For G', there was an increase in the entire temperature range with increasing NCO/OH. The same situation occurs with the loss modulus throughout the temperature range, the material with an NCO/OH of 1.05 has the lowest value. It is noted that increasing this ratio increases the value of the temperature of crossover between the modules, due to higher content of hard segments in TPU. Also, the higher modulus crossover between G' and G'' is presented by the polyurethane with the NCO/OH of 1.25. The difference between the two polyurethanes in the form of crossing is not significant.

It is expected that the sample with the highest ratio NCO/OH, - which has the highest hard segment content- present the greatest values in the storage and loss modules, and an increase in temperature and modulus of softening due to mixture of phases. As determined by IR-FTIR spectroscopy (Table 1), with increasing NCO/OH increases the degree of phase separation (DPS), is a greater mobility of the polymer chains, so it is more ordered, crystalline, and both thermal and rheological properties are improved.

NCO/OH	1,05	1,15	1,25
DPS	83,3	86,4	88,6

Table 1. Degree phase separation (DPS) for samples with different NCO/OH.

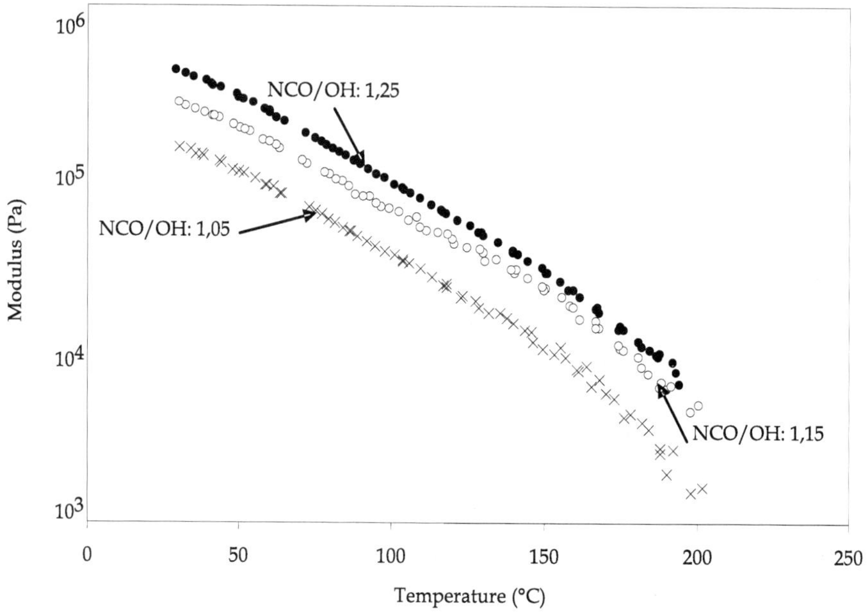

Fig. 4. Storage modulus G' for samples with different NCO/OH.

3.2 Characterization of polyurethane adhesives containing fumed silica

TPU´s with different NCO/OH and containing fumed silica were analyzed by transmission electron microscopy-TEM (Figure 5). When NCO/OH is increased, DPS in TPU´s is increased, too. Samples with fumed silica (PU105, PU115, and PU125) show an increase of DPS (Figure 6). This effect, as expected, is less evident in samples with lower NCO/OH, where there are light and dark areas, corresponding to the phases of hard and soft segments, respectively. That is, the material is less affected by the presence of fumed silica, and has the lowest phase segregation. Moreover, the degree of aggregation of silica increases with increasing the NCO/OH in the TPU´s.

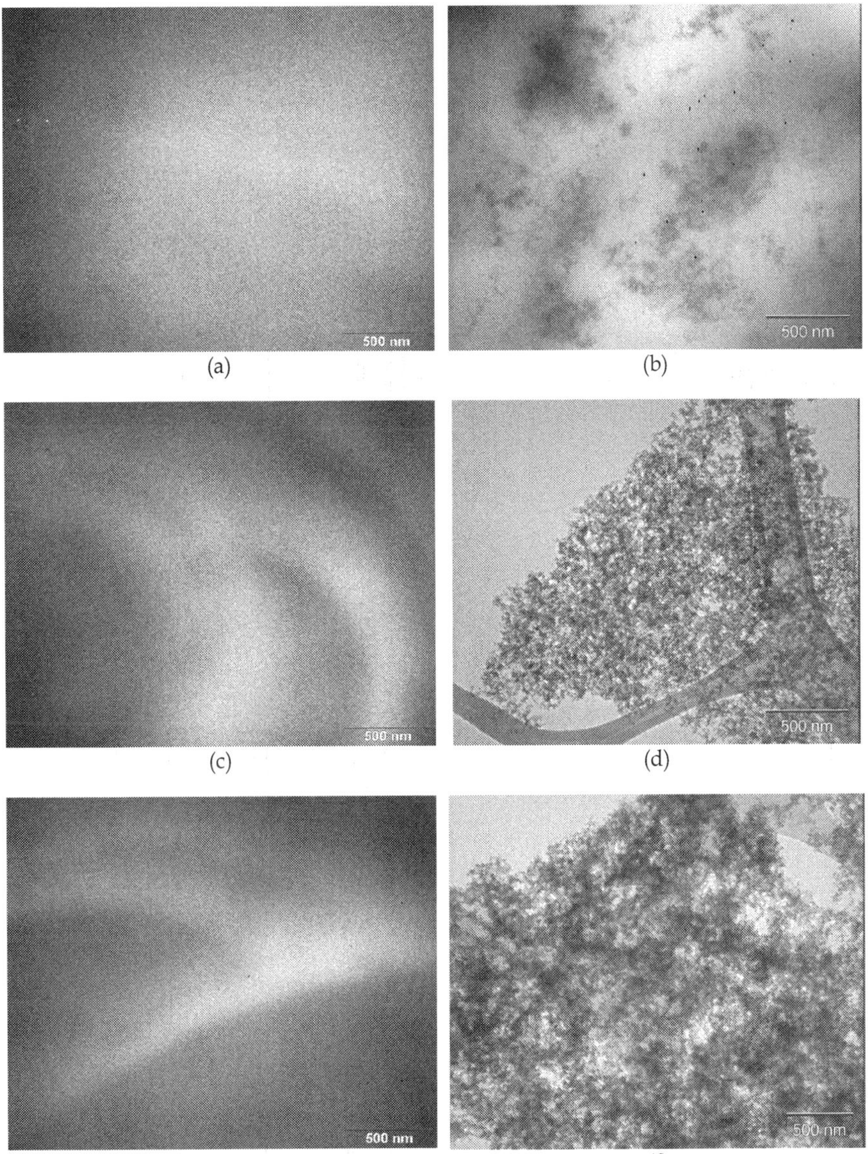

(a) (b) (c) (d) (e) (f)

(a) TPU with NCO/OH of 1,05 without silica, (b) PU105, (c) TPU with NCO/OH of 1,15 without silica, (d) PU115, (e) TPU with NCO/OH of 1,25 without silica, (f) PU125

Fig. 5. TEM of TPU with different NCO/OH.

To quantify the effect of phase segregation, we used the IR-FTIR spectroscopy. We calculated the degree of phase segregation (DPS) and the degree of phase mixing (DPM) (Torró-Palau et al., 2001, Péres-Limaña et al., 2001). The addition of fumed silica does not alter the chemical structure of TPU.

TPUS´s without silica, with increasing of NCO/OH, DPS is increased, although it increases the content of hard segments of polyurethane. With the addition of silica to polyurethane, the DPS is favored in all samples. Silanol groups increases the possibility to produce hydrogen bonds in polymer, so the links inter-urethane, more favored energetically, interact to a greater extent and promotes greater interaction between the polyurethane soft segments, resulting in increased segregation phase between hard and soft segments of TPU.

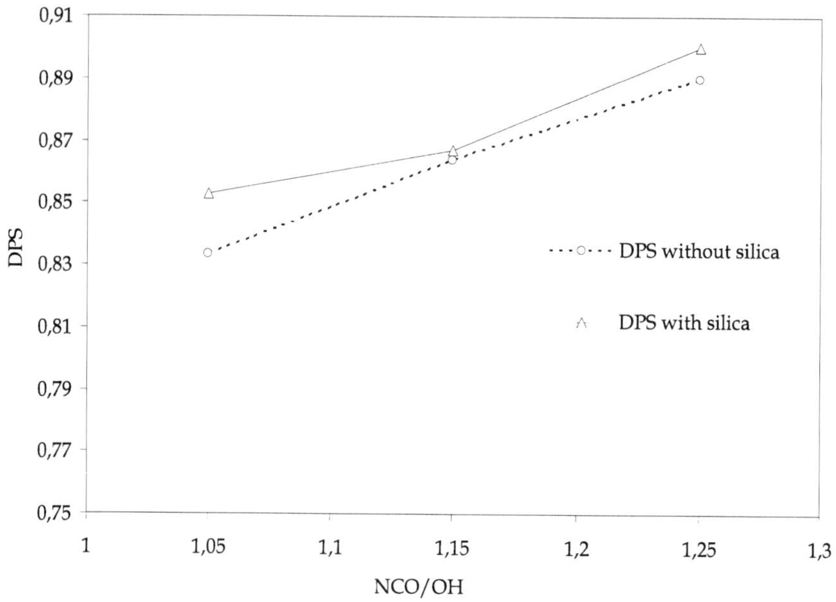

Fig. 6. Samples with and without silica with different NCO/OH

To study the interaction between TPU with different NCO/OH and silica, thermal properties and crystallinity were studied. Differential scanning calorimetry (DSC) and the X-ray diffraction were used (Figures 7 to 12).

As mentioned, the first glass transition is associated with soft segments of polyurethane. For the TPU synthesis, we used a polyol whose chains are less polar. It is expected that as the NCO/OH, increase the repulsion between hard and soft segments, and increase DPS. TPU´s will present a greater order and therefore will be more crystalline. TPU´s without silica, show an increase of T_{g1} as a response of increased in DPS (Figure 8). By incorporating fumed silica, the values of the glass transition temperature decrease over the polyurethanes do not contain silica.

In TPU´s with silica, the association-dissociation equilibrium of the hydrogen bond is favored toward the formation of more hydrogen bonds, specifically towards the formation of more interactions between the hard segments at the expense of the rupture of interactions between hard and soft segments. These interactions are stronger than interactions between soft segments themselves and silanol groups, so that TPU, despite the establishment of interactions between soft segments, they are less energetic than those between hard and soft segments, so that the polymer need less energy to reach the glass transition, crystallization or melting, these phenomena occur at lower temperatures.

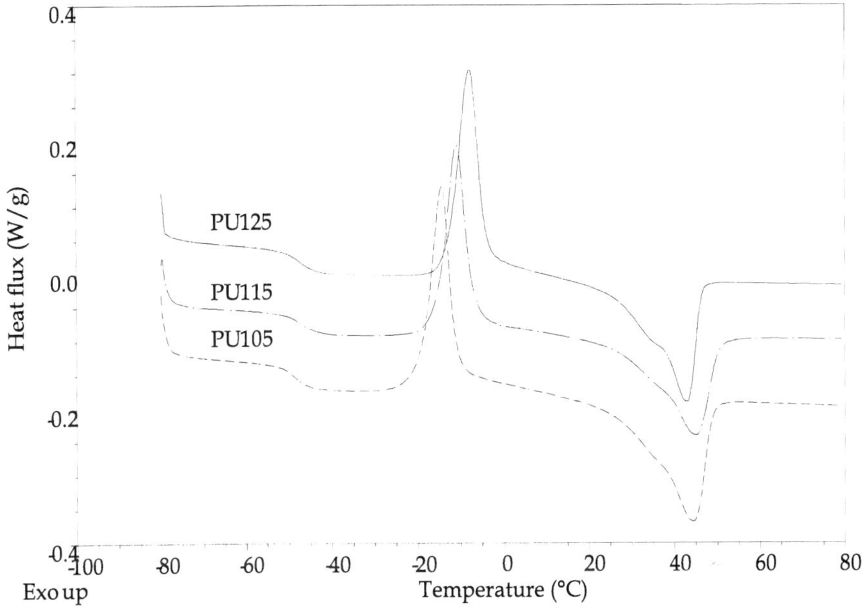

Fig. 7. DSC thermograms of synthesized thermoplastic polyurethanes (TPU) with different NCO/OH and silica.

Other properties affected by the presence of fumed silica are the enthalpy and crystallization temperature (Figures 9 and 10). During the first scan of DSC, the material is softened to 80 °C and is then rapidly cooled to -80 °C to fix the polymer chains, so that during the second sweep of temperature changes can be observed related energy with the crystal structure of the material.

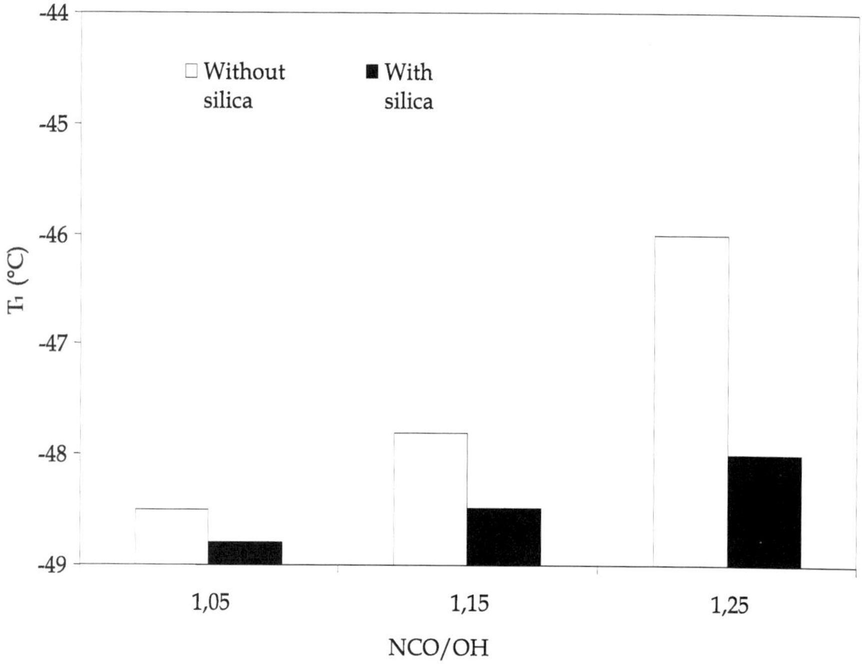

Fig. 8. Tg values of samples with different NCO/OH and with –without silica.

To compare TPU´s without silica, when NCO/OH is increased, crystallization enthalpy decreases. Also, cold crystallization process occurs at higher temperatures. PU105 without silica has highest enthalpy of crystallization, -the crystallization process is more exothermic and it occurs at a lower temperature-. In TPU, to have a lower DPS, are favored interactions between hard and soft segments, which are energetically more favorable than those observed between the soft segments themselves. By increasing the NCO/OH in TPU´s without silica, the DPS increase and establish more interactions between soft segments, which have less power than earlier, and the crystallization enthalpy decreases relative to that of PU105 without silica, and the process crystallization is observed at higher temperatures. PU125 sample without silica has therefore lower enthalpy of crystallization and the crystallization process is observed at higher temperatures.

Also, samples containing fumed silica, with increasing NCO/OH increase the enthalpy of crystallization and cold crystallization process is observed at lower temperatures. This is because the main interactions that are established in polyurethanes with silica correspond to soft segment-soft segment due to increased phase segregation with respect to TPU´s without silica, which affects the association-dissociation equilibrium of hydrogen bonding. So polyurethane sample with higher DPS is more affected by the presence of silica (PU125) and has the highest enthalpy of crystallization temperature and crystallization occurs at lower values for PU115 and PU105. Finally, we observe that TPU´s without silica, release more energy during heating process comparing with samples with silica. Also, the cold crystallization process of polyurethane is observed at a lower temperature. This is because the samples without silica have a lower DPS.

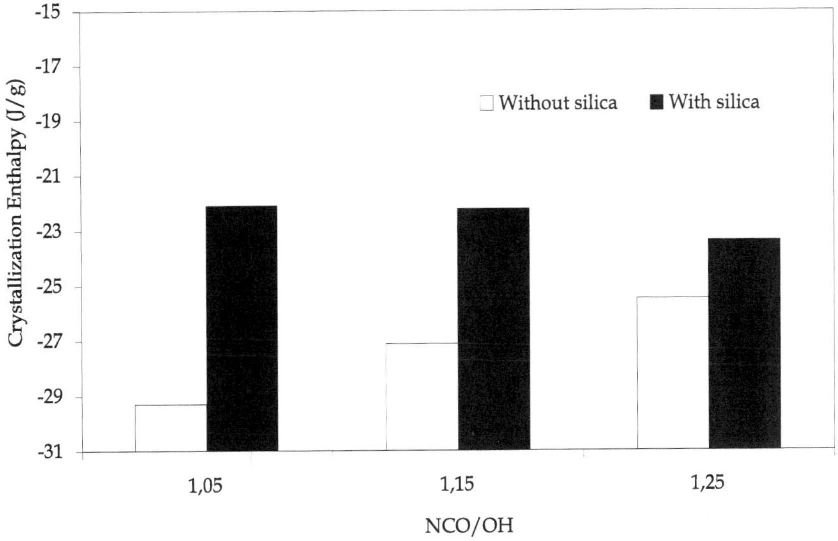

Fig. 9. Enthalpy of crystallization and cold crystallization of samples with different NCO/OH and with -without silica.

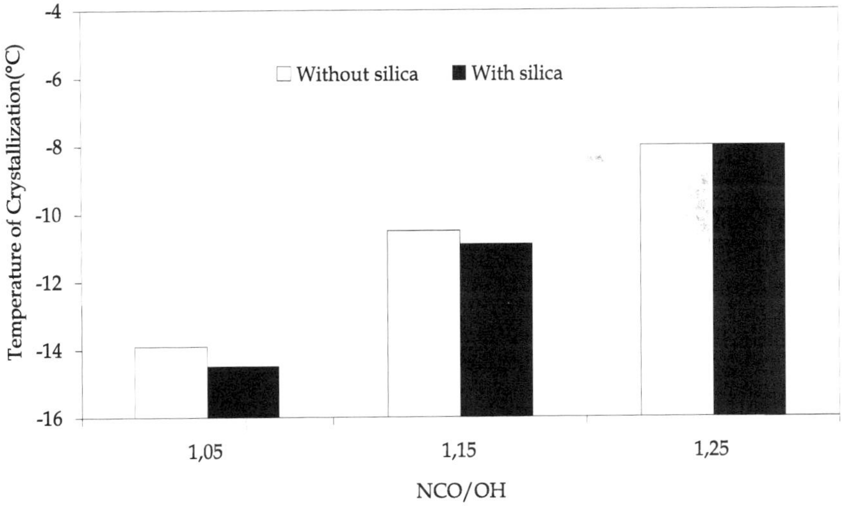

Fig. 10. Temperature of crystallization of samples with different NCO/OH and with - without silica.

During fusion enthalpy (softening) in TPU´s without silica, the melting enthalpy decreases with increasing NCO/OH, as a result of increased phase segregation of polyurethane, and it is necessary to apply a lower energy content to achieve softening of the polyurethane. PU105 without silica has the highest melting enthalpy, - process that needs more energy to soften the polymer - and as expected, there is a greater melting temperature. In TPU´s with lower DPS, are favored interactions between hard and soft segments, which are energetically more favorable than those, observed between the soft segments themselves. By increasing the NCO/OH in TPU´s without silica, DPS increase and establish more interactions between soft segments, which have less energy than before, and so the melting enthalpy decreases and the melting process occurs at lower temperatures. PU125 sample without silica has therefore lower enthalpy of fusion. In this case, the difference between the melting temperatures in TPU´s without silica with different NCO/OH is not significant.

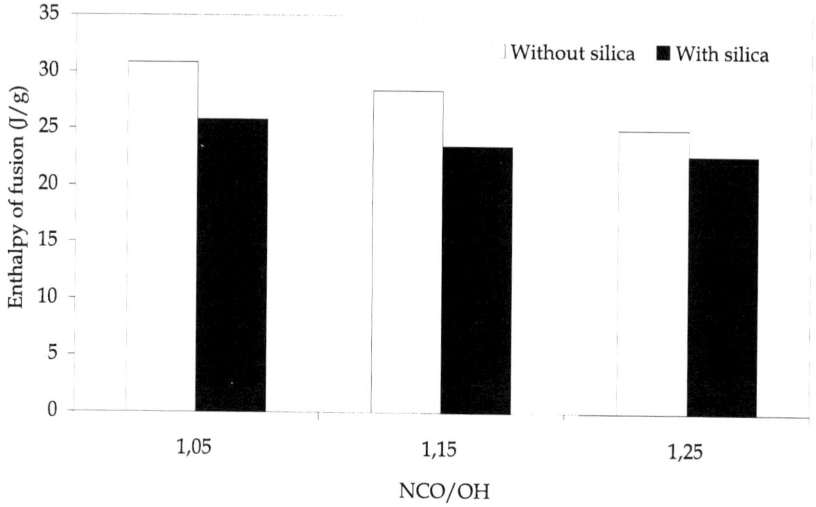

Fig. 11. Enthalpy of fusion of samples with different NCO/OH and with -without silica.

In polyurethanes containing fumed silica, with increasing NCO/OH, the melting enthalpy decreases and the softening process occurs at lower temperatures. This is because the main interactions that are established in TPU with silica correspond to the soft segments themselves, due to increased phase segregation due to the effect of the presence of silica on the association-dissociation equilibrium hydrogen bond. So, TPU with higher DPS, is more affected by the presence of silica (PU125) and has the lowest melting enthalpy and melting temperature is observed at lower values for other TPU´s with silica.

Finally, polyurethanes do not contain silica; require more energy during the heating process to melt for TPU´s with silica. Also, the merger of polyurethane without silicon is observed at a higher temperature than the samples containing hydrophilic silica. This is because the samples without silica have a lower DPS, thus favoring interactions between hard and soft segments, which are energetically stronger than those, observed between the soft segments in TPU´s themselves with silica, as required more energy for melting.

Also, It was used X-ray diffraction (XRD). Results show -as in previous studies-, significant diffraction peaks at 2θ = 20° and 2θ = 25° (without silica). TPU´s with silica present three main reflections at 2θ: (21.2° - 21.7°), 2θ: (22.2° - 22.4°) and 2θ (24.1° to 24.6°). Some reflections of (101) are insignificant, so it was not possible to quantify.

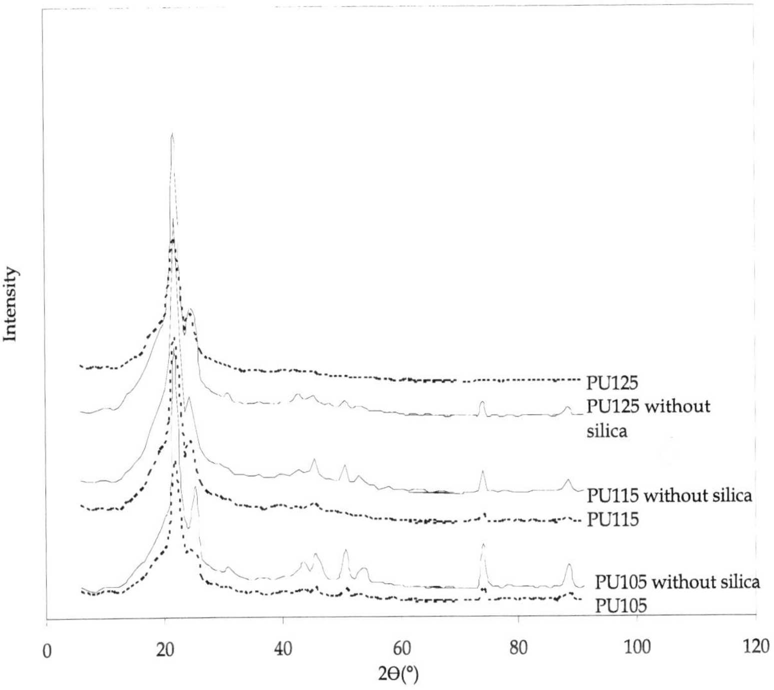

Fig. 12. X-ray diffraction (XRD) of samples with different NCO/OH and with –without silica.

4. Conclusions

The morphological study of polyurethanes without silica indicates that the increase of NCO / OH increases the degree of phase segregation (DPS), due to the effect of repulsion that exists between the polar hard segments of polyurethane and non-polar chains polyol. With the addition of hydrophilic silica to polyurethane, the degree of phase separation is favored in all polyurethanes, indicating a possible interaction of silica silanol groups by hydrogen bonds with the polymer.

5. References

Jaúregui-Beloqui, B., Fernández-García, J.C., Orgilés-Barceló, A.C., Mahiques-Bujanda, M.M. & Martín-Martínez, J.M. (1999). Thermoplastic polyurethane-fumed silica composites: influence of the specific surface area of fumed silica on the viscoelastic and adhesion properties. *Journal of Adhesion Science and Technology*, 13, pp. 695-711, 0169-4243.

Jaúregui-Beloqui, B., Fernández-García, J.C., Orgilés-Barceló, A.C., Mahiques-Bujanda, M.M. & Martín-Martínez, J.M. (1999). Rheological properties of thermoplastic polyurethane adhesive solutions containing fumed silicas of different surface areas. *International Journal of Adhesion and Adhesives*, 19, pp. 321-328, 0143-7496.

Maciá-Agulló, T.G., Fernández-García, J.C., Pastor-Sempere, N., Orgilés-Barceló, A.C. & Martín-Martínez. J.M. (1992). Addition of Silica to Polyurethane Adhesives. Journal of Adhesion, 38, pp. 31-53, 0021-8464.

Navarro-Bañón, V., Vega-Baudrit, J., Vázquez, P. & Martín-Martínez, J.M. (2005). Interactions in Nanosilica-Polyurethane Composites Evidenced by Plate-Plate Rheology and DMTA. *Macromolecular Symposia*, 221, pp. 1, 1022-1360.

Nunes, R.C.R., Fonseca, J.L.C.M. & Pereira, M.R. (2000). Polymer–filler interactions and mechanical properties of a polyurethane elastomer. (2000). *Polymer Testing*, 19, pp. 93-103, 0142-9418.

Nunes, R.C.R., Pereira, R.A., Fonseca, J.L.C. & Pereira, M.R. (2001). X-ray studies on compositions of polyurethane and silica. *Polymer Testing*, 20, pp. 707-712, 0142-9418.

Oertel, G. (1993). *Polyurethane Handbook 2nd*. Hanser: New York, pp. 7-116. 3-446-17198-3.

Pérez-Limiñana, M.A., Torró-Palau, A.M., Orgilés-Barceló, A.C. & Martín-Martínez, J.M. (2001). Rheological properties of polyurethanes adhesives containing silica as filler; influence of the nature and surface chemistry of silica. *Macromolecular Symposia*, 169, pp. 191-196, 1022-1360.

Sánchez-Adsuar, M.S., Papón, E. & Villenave, J. (2000). Influence of the prepolymerization on the properties of thermoplastic polyurethane elastomers. Part I. Prepolimerization characterization. *Journañ of Applied Polymer Science*, 76, pp. 1596-1601, 0021-8995.

Tien Y. & Wei K. (2001). Hydrogen bonding and mechanical properties in segmented montmorillonite/polyuretane nanocomposites of different hard segment ratios. *Polymer*, 42, pp. 3213-3221, 0032-3861.

Torró-Palau, A., Fernández-García, J.C., Orgilés-Barceló, A.C. & Martín-Martínez, J.M. (2001). Characterization of polyurethanes containing different silicas. *International Journal of Adhesion and Adhesives*, 21, pp. 1-9, 0143-7496.

Vega-Baudrit, J., Navarro-Bañon, V., Vázquez, P. & Martín-Martínez, J.M. (2006). Properties of thermoplastic polyurethane adhesives containing nanosilicas with different specific surface area and silanol content. *International Journal of Adhesion and Adhesives*, 27, pp. 469-479, 0143-7496.

Vega-Baudrit, J., Sibaja-Ballestero, M., Vázquez, P., Navarro-Bañón, V., Martín-Martínez, J.M. & Benavides, L. (2005). Kinetics of Isothermal Degradation Studies in Adhesives by Thermogravimetric Data: Effect of Hydrophilic Nanosilica Fillers on the Thermal Properties of Thermoplastic Polyurethane-Silica Nanocomposites. *Recent Patents on Nanotechnology*, 2(3), pp. 220-226, 1872-2105.

Vega-Baudrit, J., Sibaja-Ballestero, M. & Martín-Martínez, J.M. (2009). Study of the Relationship between Nanoparticles of Silica and Thermoplastic Polymer (TPU) in Nanocomposites. *Journal of Nanotechnology Progress International. (JONPI)*, 1, pp. 24-34, 1941-3475.

3

Manufacture of Different Types of Thermoplastic

Lavinia Ardelean, Cristina Bortun, Angela Podariu and Laura Rusu
"Victor Babes" University of Medicine and Pharmacy Timisoara
Romania

1. Introduction

The development of resins represented a great step forward in dental technique, the first thermopolymerisable acrylic resins being developed in 1936. Acrylic resins are better known as poly(methyl methacrylate) or PMMA. They are synthetically obtained materials that can be modelled, packed or injected into molds during an initial plastic phase which solidify through a chemical reaction-polymerisation (Phoenix et al., 2004). However, the disadvantages of thermopolymerisable acrylic resins, connected to increased porosity, high water retention, volume variations and irritating effect of the residual monomer (organic solvent, hepatotoxic), awkward wrapping system, difficult processing, together with the polymer development, have led to alternative materials such as polyamides (nylon), acetal resins, epoxy resins, polystyrene, polycarbonate resins etc. (Negrutiu et al., 2001).

Thermoplastic resins have been used in dental medicine for fifty years. In the meantime, their use has spread due to their superior characteristics. Their ongoing development has yielded new classes of more and more advanced materials and technologies, which make possible the manufacturing of dentures with better splinting properties then traditional dentures.

2. Thermoplastic resins used in dentistry

The classification of resins according to DIN EN ISO–1567 comprises:

From the point of view of their composition, as far as thermoplastic resins are concerned, we can distinguish among: acetal resins, polycarbonate resins (belonging to the group of polyester resins), acrylic resins, polyamides (nylons).

Usage of thermoplastic resins in dental medicine has significantly grown in the last decade. The technology is based on plasticising the material using only thermal processing in the absence of any chemical reaction. The possibility of injecting the plasticized resin into a mold has opened a new perspective to full denture and removable partial denture technology.

Successive alterations to the chemical composition led to the diversification of their range of application, so that at present thermoplastic materials are suitable for the manufacturing of removable partial dentures which totally or partially eliminate the metallic component, resulting in the so-called "metal-free removable partial dentures" (Bortun et al., 2006).

Type	Class (manufacturing)	Group (presentation form)
Type 1	thermopolymerisable resins (> 65°C)	Groupe 1: bicomponent - powder and liquid Groupe 2: monocomponent
Type 2	autopolymerisable resins (< 65°C)	Groupe 1: bicomponent - powder and liquid Groupe 2: bicomponent - powder and casting liquid
Type 3	thermoplastic resins	Monocomponent system: grains in cartridges
Type 4	photopolymerisable resins	Monocomponent system
Type 5	microwave polymerisable resins	Bicomponent system

Table 1. The classification of resins according to DIN EN ISO-1567

Indications for thermoplastic resins include: partial dentures, preformed clasps, partial denture frameworks, temporary or provisional crowns and bridges, full dentures, orthodontic appliances, myofunctional therapy devices, anti-snoring devices, different types of mouthguards and splints.

2.1 Thermoplastic acetal

Thermoplastic acetal is a poly(oxy-methylene)-based material, which as a homopolymer has good short-term mechanical properties, but as a copolymer has better long-term stability (Arikan et al., 2005).

Acetal resin is very strong, resists wear and fracturing, and it's flexible, which makes it an ideal material for pre-formed clasps for partial dentures, single pressed unilateral partial dentures, partial denture frameworks, provisional bridges, occlusal splints and implant abutments, partial denture frameworks, artificial teeth for removable dentures, orthodontic appliances.

Acetal resins resist occlusal wear and are well suited for maintaining vertical dimension during provisional restorative therapy. Acetal does not have the natural translucency and esthetic appearance of thermoplastic acrylic and polycarbonate (Ozkan et al., 2005).

2.2 Thermoplastic polyamide (nylon)

Thermoplastic nylon is a polyamidic resin derived from diamine and dibasic acid monomers. Nylon is a versatile material, suitable for a broad range of applications.

Nylon exhibits high flexibility, physical strength, heat and chemical resistance. It can be easily modified to increase stiffness and wear resistance. Because of its excellent balance of strength, ductility and heat resistance, nylon is an outstanding candidate for metal replacement applications.

They are used primarily for tissue supported removable dentures because their stiffness makes them unsuitable for usage as occlusal rests or denture parts that need to be rigid. Because of its flexibility, it can't maintain vertical dimension when used in direct occlusal forces.

Nylon is a little more difficult to adjust and polish, but the resin can be semi-translucent and provides excellent esthetics (Donovan & Cho, 2003).

Resin type	Main substance	Resistance	Durity	Flexibility	Esthetics	Biocompatibility
Acetalic resin	Polioxime tylen	very good	very high	medium	good	very good
Polyamidic resin	diamin	good	high	medium or very high, depending on the material	very good	very good

Table 2. Comparative aspects of acetalic and polyamidic thermoplastic resins

2.3 Thermoplastic polyester

Another group of thermoplastic materials used in dentistry are polyester resins. These resins melt between 230-290°C and the technology implies casting into molds.

Polycarbonate resins are particular polyester materials. They exhibit fracture strength and flexibility, but the wear resistance is lower when compared to acetal resins. However, polycarbonates have a natural translucency and finishes very well, which make them proper for producing temporary restorations. They are not suitable for partial denture frameworks (Negrutiu et al., 2005).

At present, there are several manufacturers that provide thermoplastic materials for dental use: The Flexite Company, Valplast Int. Corp., Girrbach Dental, Bredent, Dentsply, DR Dental Resource Inc., If Dental-Pressing Dental etc.

(a)

(b)

(c)

Fig. 1. (a), (c) Cartridges of different thermoplastic resins, (b) The granular aspect of the material

2.4 Presentation form and injection

Thermoplastic materials can be polymerised or prepolymerised and they can be found in granular form, with low molecular weight, already wrapped in cartridges which eliminates dosage errors - Fig. 1.

They have a low plasticizing temperature and exhibit a high rigidity in spite of their low molecular weight. Their plasticizing temperature is 200-250°C.

After thermal plasticization in special devices, the material is injected under pressure into a mold, without any chemical reactions. The metallic cartridges containing thermoplastic grains are heated to plasticize the resin. The cartridges are set in place into the injecting unit and pressure of 6-8 barrs is used to force the plasticized resin to fill the mold. Pressure, temperature and injecting time are automatically controlled by the injecting unit. This results in compact dentures with excellent esthetics and good compatibility.

Injecting thermoplastic resins into molds is not a common technology in dental laboratories because the need of expensive equipment and this could be a disadvantage.

We will describe the manufacturing process of metal-free removable partial dentures made off several thermoplastic resins in different cases of partial edentations, with removable partial dentures without metallic frame, or combining the metallic frame with thermoplastic resin saddles, selected according to the requirements of the indications and manufacturing technology- Fig. 2.

(a)

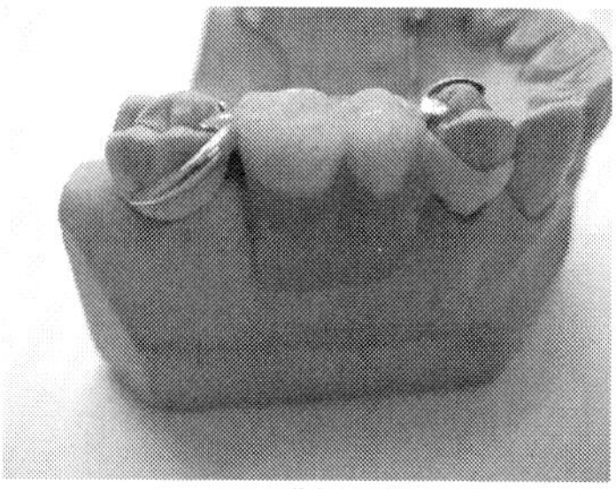

(b)

(c)

Fig. 2. Different combinations between thermoplastic resins. (a), (c) Without metal, (b) With metal

The main characteristics of thermoplastic resins used are: they are monomer-free and consequently non-toxic and non-allergenic, they are injected by using special devices, they are biocompatible, they have enhanced esthetics and are comfortable at wearing.

The special injection devices we use are Polyapress (Bredent) and R-3C (Flexite) injectors- Fig. 3.

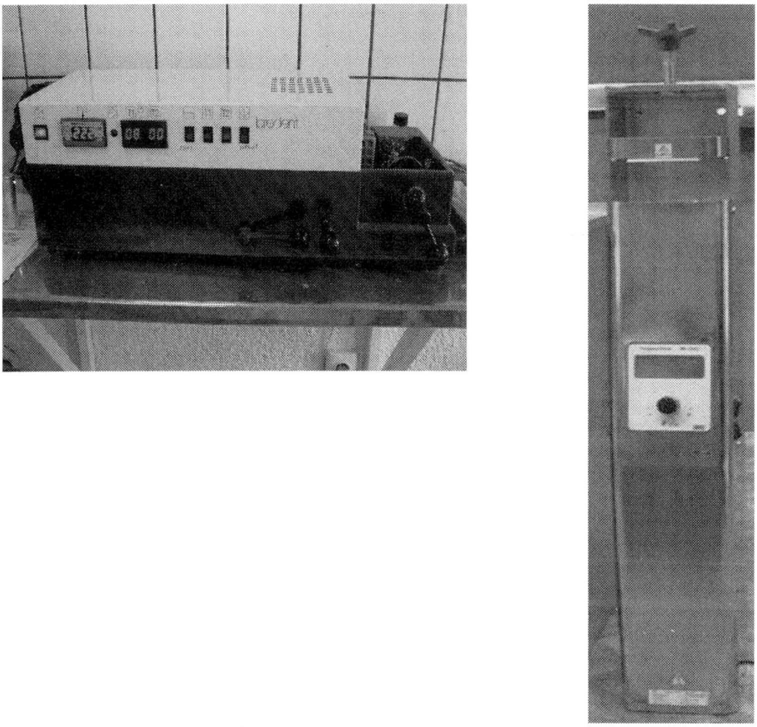

(a) (b)

Fig. 3. (a) The Polyapress injection-molding device (Bredent), (b) The R-3C injector (Flexite)

3. Manufacture technology for acetal-resin dentures

The acetal resin has optimal physical and chemical properties and it is indicated in making frames and clasps for removable partial dentures, being available in tooth colour and in pink.

The denture acetal resin framework was combined with the use of acrylic resins at saddle level (Fig. 2). As a particularity of the manufacturing we mention the fact that it is necessary to oversize the main connector, clasps and spurs, because the resistance values characteristic for the acetal resin do not reach those of a metal framework. Injection was carried out using the R-3 C digital control device that has five preset programmes, as well as programmes that can be individually set by the user.

The maintenance, support and stabilizing systems used are those with metal-free, Ackers circular clasps, chosen according to the median line of the abutment teeth and the insertion axis of the denture.

The significant aspects of the technical steps in the technology of removable partial dentures made of thermoplastic materials are described.

3.1 The working model

The working model is poured of class IV hard plaster, using a vibrating table, in two copies (Fig. 4), as one of the models gets deteriorated when the acetal component of the denture is dismantled.

Fig. 4. Casting the working model

3.2 Parallelograph analysis and framework design

The model is analyzed by parallelograph in order to assess its retentiveness and to determine the place where the active arms of the clasp are placed-Fig. 5.

The abutment teeth were selected and the position of the cast was chosen and recorded so that a favourable path of insertion was obtained.

Tripod marks were used to record the position of the cast. Carbon graphite rod was used to mark the heights of contour on the abutment teeth and the retentive muco-osseous tissues. Undercut gauges were used to measure the abutments undercuts. Engagement of the terminal third of the retentive arms of the clasps was established at 0.25 mm below the greatest convexities for each abutment.

After the parallelograph analysis was carried out, a soft tip black pencil was used to draw the future framework design on the model. The design included all extensions of saddles, major connector, retentive and bracing arms of the clasps, occlusal rests and minor connectors of Akers circumferential clasps on abutment teeth.

The design starts with the saddles, following the main connector, the retentive and opposing clasp arms, the spurs and secondary connectors of the Ackers circular clasps.

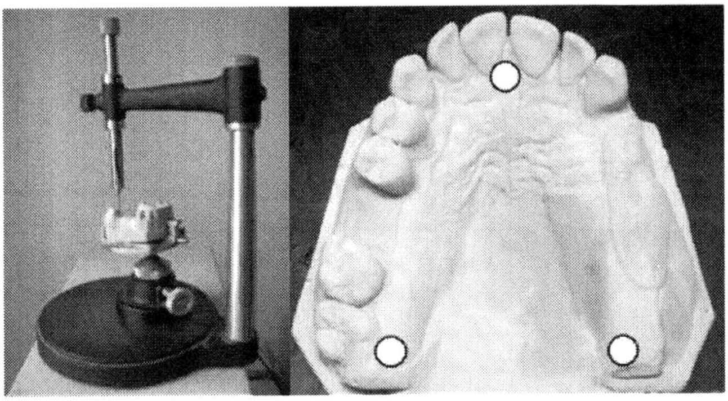

Fig. 5. Parallelograph analysis

3.3 Duplication of the master model

After designing the framework, the master model is prepared for duplication, including foliation and deretentivisation (Fig. 6). At the beginning, blue wax plates are used as spacers in regions where the framework has to be spaced from the gingival tissue. The residual ridges are covered with 1 mm thick wax along the 2/3 of the mesio-distal length and the 2/3 of the lingual slope height. The wax crosses the edge of the ridge and also covers a short portion of the buccal slope. The same thickness of the spacer is used along the mucosal region of the major connector where the wax is applied between the gingival margin and the bottom of the alveolo-lingual sulcus. A 0.3 mm wax spacer is placed along the place of minor connectors.

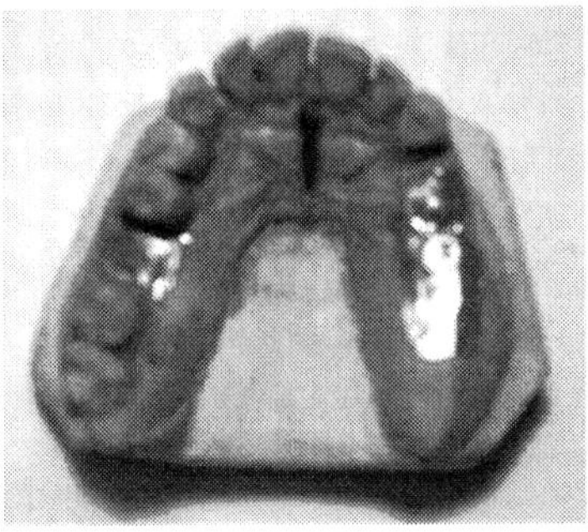

Fig. 6. Foliation and deretentivisation of the model

The next step is the block-out procedure. Block-out wax is applied between teeth cervices and gingival margin of the drawing representing the clasps arms. A smooth joint is made between block-out wax and spacing wax.

Duplication of the master cast is done in the usual manner, using a vinyl-polysiloxane silicone placed in a conformer. After the silicone bounds, the impression is taken and the duplicate model is cast (Fig. 7), using class IV hard plaster.

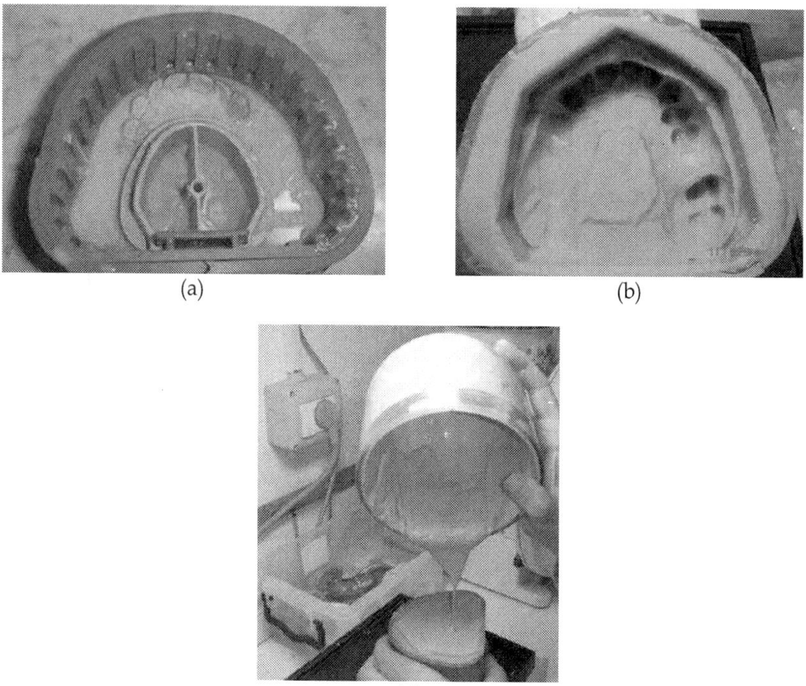

(a) (b) (c)

Fig. 7. Duplication of the model: (a) Conformer with the model placed inside, (b) Silicone impression, (c) Casting of the working model using class IV hard plaster

3.4 Wax pattern manufacturing

The wax pattern of the removable partial denture is manufactured following the profiles imprinted on the model (Fig. 8): the wax pattern of the main connector, made of red wax (so that it's thickness is twice as normal), the wax pattern of the saddles and the wax pattern of the Ackers circular clasps, made of blue wax.

Injection bars are required for the sensitive areas of the framework that are placed on the areas that are not visible in the finite piece.

A large central shaft is also necessary in order to connect with the main connector, through which the initial injection takes place.

Unlike the pattern of a metallic framework, the patterns of the clasps, occlusal rests and lingual bar were made 50% thicker.

Because the wax pattern of the metal-free framework has to be 50% thicker than that of a metallic framework, pink wax is used for wax-up. In order to produce patterns of the saddles, wax plates were adapted on the cast according to the hallmarks and circular retentive holes were cut along them.

The lingual bar was made by the same wax, achieving a half-pear shape with an optimal dimension. Wax-up of the saddles and lingual bar was made using a special wax, easy to wash away, following the hallmarks. Preformed wax patterns were adapted to the hallmarks with an adhesive solution. Blue wax was used to drop wax-up the patterns of the circumferential clasps.

Once the pattern of the framework is ready, it is stabilized by sticking the margins to the cast.

3.5 Investing the wax pattern

Spruing the framework was performed using five minor sprues of 2.5 mm calibrated wax connected to one major sprue.

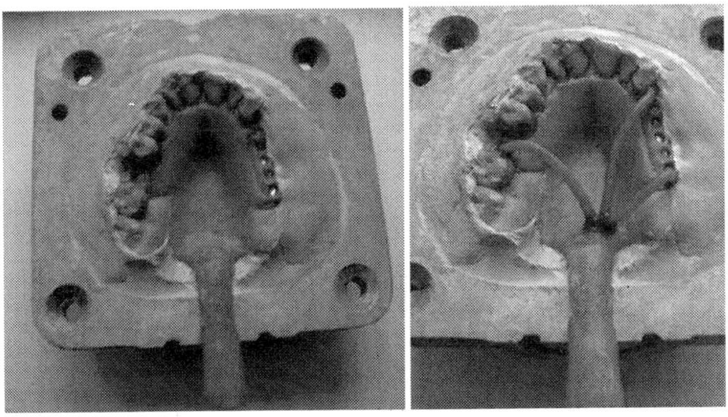

Fig. 8. Wrapping the wax pattern frame of the removable partial denture

After surface-tension reducing solution is applied to the wax pattern, it is invested in a vaseline insulated aluminum flask. Class III hard stone is used as investment. About 250 g gypsum paste is poured into one of the two halves of the flask and the duplicated cast containing the spruing of the framework pattern is centrally dipped base-face down-Fig. 9.

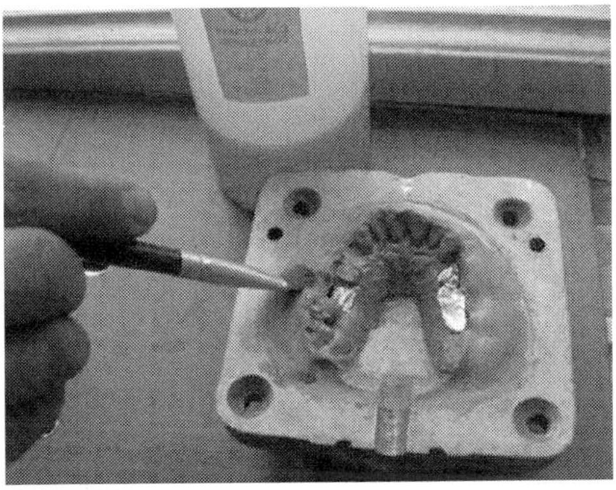

Fig. 9. Insulation of the investment

When the investment is set, the gypsum surface is insulated and the second half of the flask is assembled. About 400 g of the same hard stone is prepared and poured into the upper chamber of the flask, covering thoroughly the wax pattern and sprues.

After the gypsum sets the flask is submerged in warm water in a thermostatic container.

The two halves of the flask are than disassembled and the wax is boiled out using clean hot water.

The mold is then insulated using a special agent which is applied in a single layer on the gypsum surface. The surface of the mold is given a shining aspect by treating the gypsum surface with light curing transparent varnish.

3.6 Injection of the thermoplastic acetal resin framework

Injection is carried out with the R-3C (Flexite) injector-Fig. 10 a, which does not take up much space as it can be mounted on a wall as well.

The device has the following parameters: digital control, preset programmes for: Flexite Plus, Flexite Supreme and Flexite MP, Northerm, Proguard and programmes that can be individually set by the user. The pressure developed is 6-8 barrs.

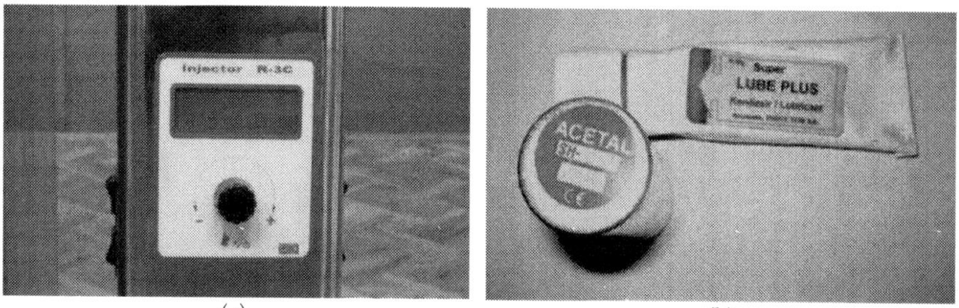

(a) (b)

Fig. 10. (a) R-3C injector (Flexite), (b) The flask with grains of thermoplastic material and the lubricant used

Before the injection procedure, the valves of carbon dioxide tank are checked to make sure the injecting pressure was according to procedure demands (7.2-7.5 barrs). Preheating temperature and time are also checked (15 minutes at 220°C).

The corresponding cartridge of injecting material (quantity and color) is selected - Fig. 10 b. The cartridge is introduced into one of the two heating cylinders after a vaseline base lubricant has been applied at its closed end - Fig. 10 b. The cartridge membrane is pointed to the flask chamber.

The excess of silicone vaseline lubricant on the margin of the heating cylinder is wiped out using a highly absorbent paper.

Preheating process is then activated by pushing the key on the front control panel-Fig 11. When the programmed preheating time elapses, an audible signal is heard.

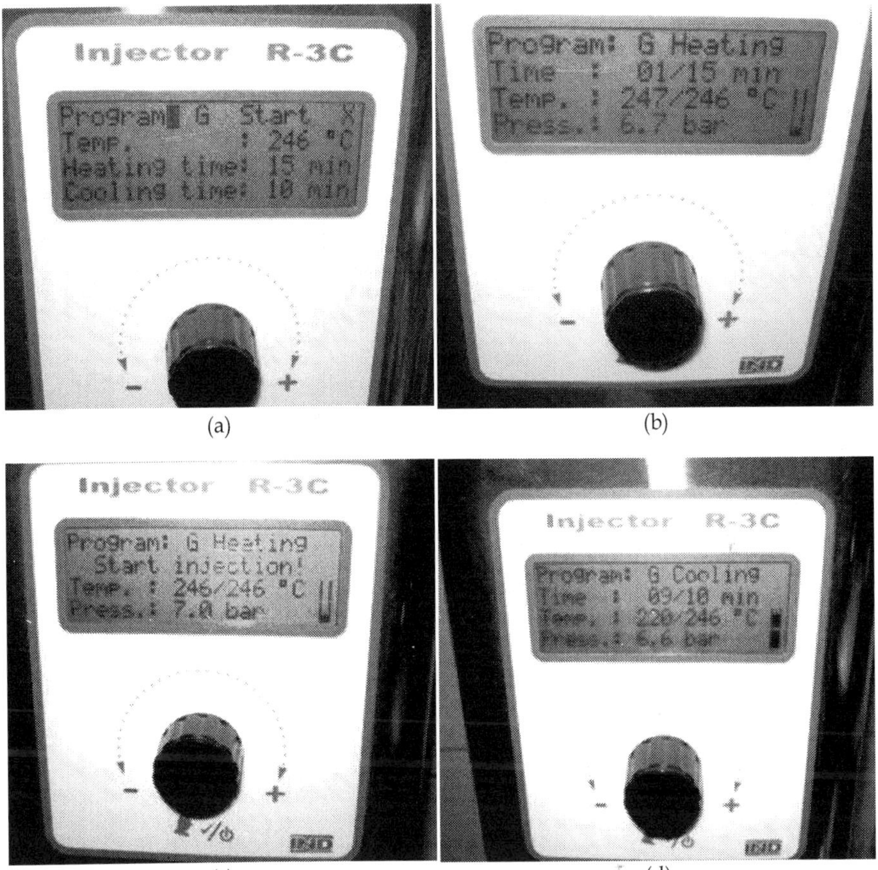

Fig. 11. Schedule of „G" program of injecting the thermoplastic material: (a) Start, (b) Heating, (c) Injecting, (d) Cooling

The two halves of the flask are assembled and fastened with screws. If the flask has been assembled earlier, water vapor condensation might have occurred inside the mold, which would have had a negative effect on the quality of the injected material.

The flask is inserted and secured in the corresponding place of the injecting unit. The opening of the flask is set in a straight line with the heating cylinder and cartridge.

The heating cylinder containing the material cartridge is brought near the flask and the injecting procedure is initiated by pressing the key on the control panel. The injection process takes 0.25 seconds. The pressure is automatically kept constant for one minute so that setting contraction is compensated. This stage is indicated with the sign "----" on the screen.

The cylinder is than moved about 3 mm away from the flask so that the cartridge could be separated using a trowel and a mallet. The flask is then released and pulled out. The used cartridge is automatically pushed out pressing the evacuation key.

In order to achieve optimal quality of the material, the flask is left to cool slowly for 8 hours.

3.7 Disassembling and finishing the acetal framework

Before investment removal, screws are loosened and the flask is gently disassembled.

The stone blocking the vents in the upper side of the flask are removed using a hook and a mallet- Fig. 12.

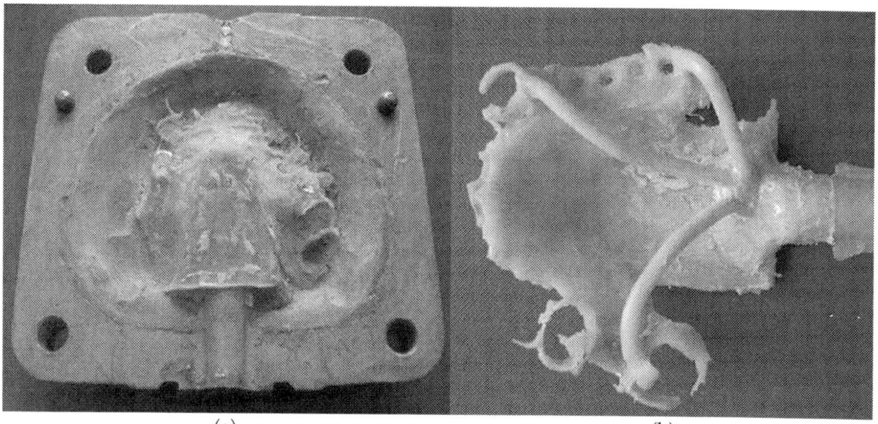

(a) (b)

Fig. 12. Disassembling the framework of the acetal resin removable partial denture. (a) The framework is still in the flask, (b) Disassembling is complete

Any excess of vaseline in the injecting canal is removed so that the injected material wouldn't contain any such remains during subsequent usage.

The sprues are cut off using low-pressure carbide and diamond burs to avoid overheating the material.

Finishing and polishing was performed using soft brushes, ragwheel and polishing paste - Fig. 13.

Disassembling the frame of the future removable partial denture is followed by matching it to the model, processing and finishing this component of the framework denture - Fig. 14.

(a)

(b)

(c)

Fig. 13. (a) Tools used for processing the acetal framework, (b) Tools used for finishing and polishing the acetal framework, (c) Special polishing paste

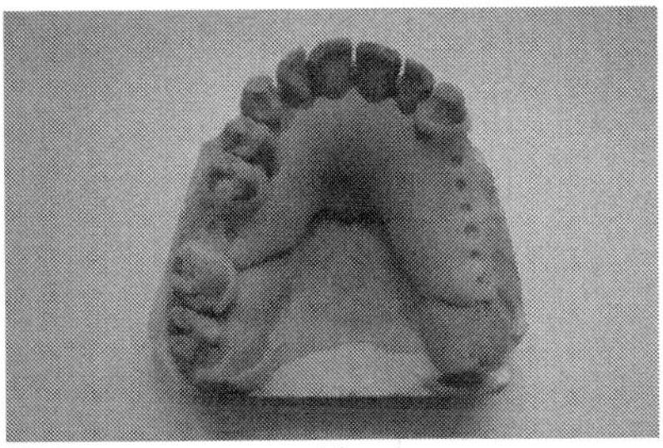

(a)

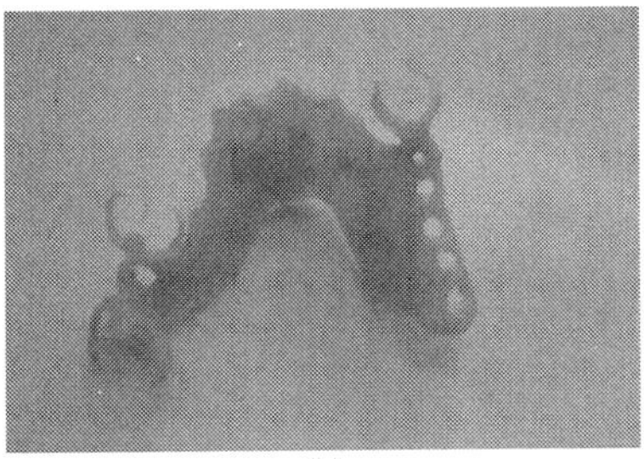

(b)

Fig. 14. (a) Matching the acetal framework to the model, (b) The finished acetal framework

3.8 Saddle manufacturing and teeth mounting

Once the framework is ready, the artificial teeth are set up. Wax patterns of the saddles are constructed by dropping pink wax over the framework, set in place on the master model. The wax is extended to the bottom of the buccal and alveololingual sulci. Teeth set-up starts with the most mesial tooth, which is polished until it esthetically fits onto the arch.

When all the teeth are properly set, an investing procedure is used to turn the wax pattern into acrylic saddles. Putty condensation silicone is used to make an impression of the wax pattern placed on the master model. When silicone is set, impression is detached, wax removed, and teeth, framework and the master model are thoroughly cleaned. Openings are being cut on the lateral sides of the impression and the teeth are set in the corresponding places inside the impression. The master model is insulated. The framework is placed on the model and the impression set in its original place. The acrylic component of the denture is wrapped according to traditional methods, using rectangular flasks in which the wax pattern is embedded into class II plaster (Fig. 15 a).

Self curing acrylic resin is prepared and poured inside the impression through the lateral openings. The cast is introduced into a heat-pressure curing unit setting a temperature of 50°C and a pressure of 6 barrs for 10 minutes to avoid bubble development. Once the resin is cured, the impression is removed. Burs, brushes, ragwheels and pumice are used to remove the excess, polish and finish the removable partial denture (Fig. 15 b).

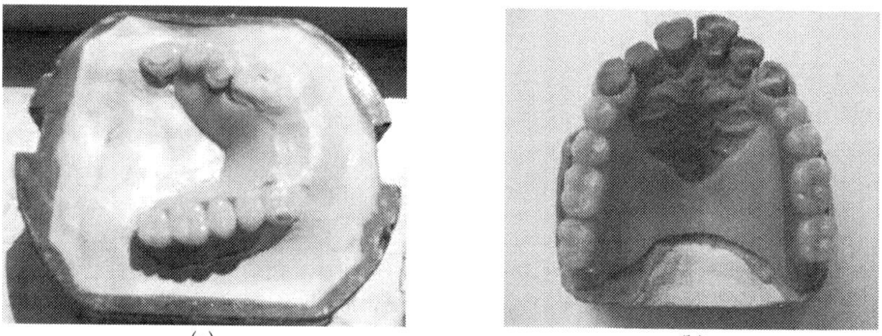

(a) (b)

Fig. 15. (a) Wrapped wax pattern with teeth, (b) Partial dentures made of acetal resin and acrylic resin

The result is a consistent removable partial denture with no macroscopic deficiency even in the thinnest 0.3-0.5 mm areas of clasps, which means the technology is effective.

4. Splints made of acetal resin

Due to the fact that among the indications of thermoplastic resins are anti-snoring devices, different types of mouthguards and splints, we experimentally manufactured acetalic resin splints (Fig. 16), in order to immobilise parodonthotic teeth, after surgery, although this is not one of the main indications for the material. Due to the fact that it matches the colour of the teeth, the splint represents a temporary postoperative esthetic solution.

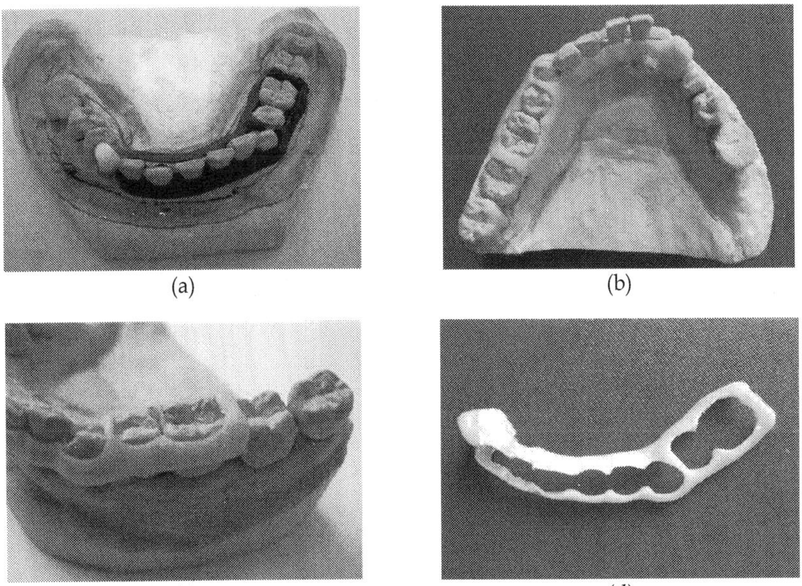

(a) (b)

(c) (d)

Fig. 16. The splint: (a) Wax pattern, (b) Thermoplastic acetal splint on the model, (c) Immobilisation splint made of acetal resin (d) Splint detached from the model

5. Acetal Kemeny-type dentures

As an experiment, in order to test the physiognomic aspect, we managed partial reduced edentations with Kemeny-type dentures (Fig. 17, 18), as an alternative to fixed partial dentures, having the advantage of a minimal loss of hard dental substance, located only at the level of the occlusal rims.

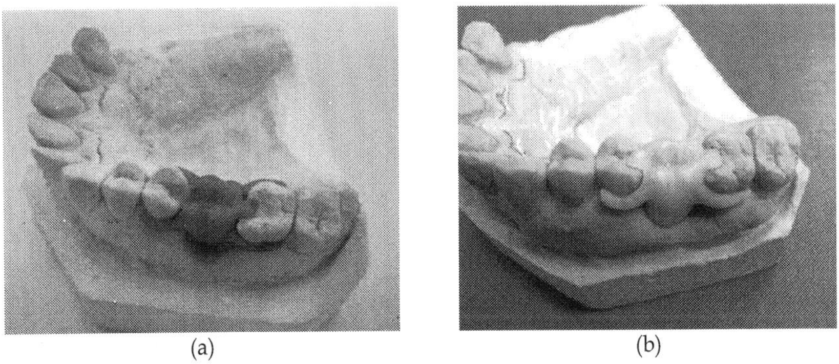

(a) (b)

Fig. 17. Kemeny dentures: (a) Unimolar denture wax pattern, (b) Denture made of acetal resin

Fig. 17 shows the way in which a molar unidental edentation can be managed, while Fig. 18 shows wax patterning aspects and manufacturing of a frontal bidental Kemeny denture made of acetal resin. The effectiveness of the technology is ensured by making artificial teeth of the same material.

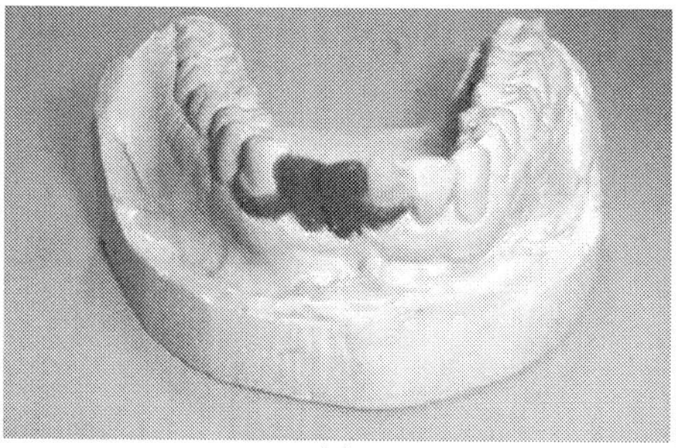

(a)

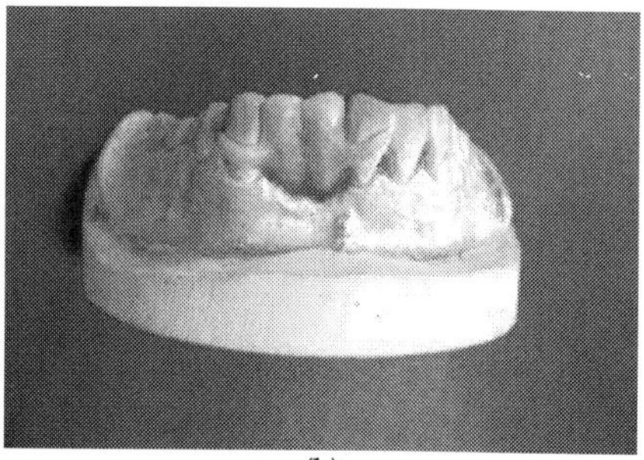

(b)

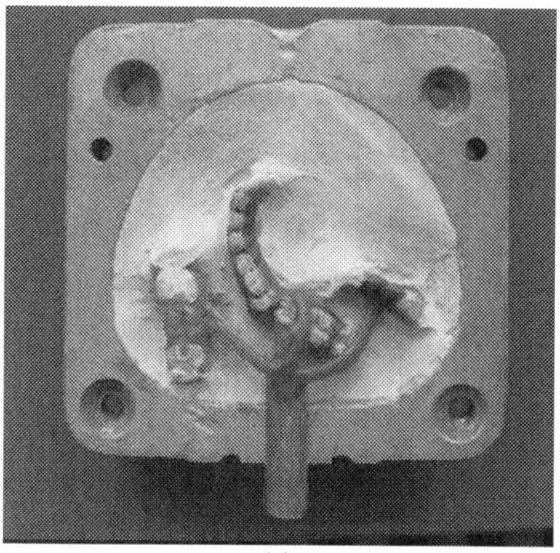

(c)

Fig. 18. (a) Frontal removable denture wax pattern, (b) Kemeny-type frontal denture, (c) Wax-patterns of the splint and Kemeny-type dentures investment system in special flask

As the material is not translucent, it is mainly suitable for dealing with lateral edentations. It can, however, be used temporarily, in the frontal area as well, in those clinical cases where short-term esthetic aspect is irrelevant.

6. Manufacture technology for polyamide resins dentures

Polyamide resin removable partial dentures are easier to make than those made of acetal resins as they do not require so many intermediary steps.

The steps are similar to those followed for acrylic dentures, differences lying in the fact that with thermoplastic materials the injecting procedure is used, and the clasps are made of the same material as the denture base, when using superflexible polyamide or we used ready-made clasps, in the case of using medium-low flexibility polyamide - Fig. 19.

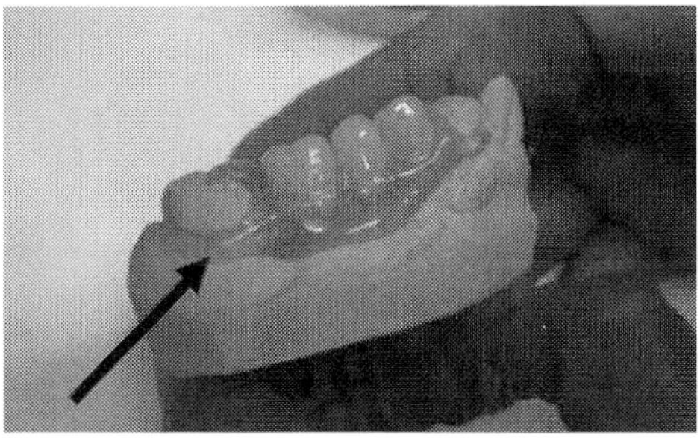

Fig. 19. Removable partial denture made of a medium-low flexibile polyamidic resin with pre-formed clasps

Using of flexible polyamide is extremely useful in cases of retentive dental fields- Fig. 20.

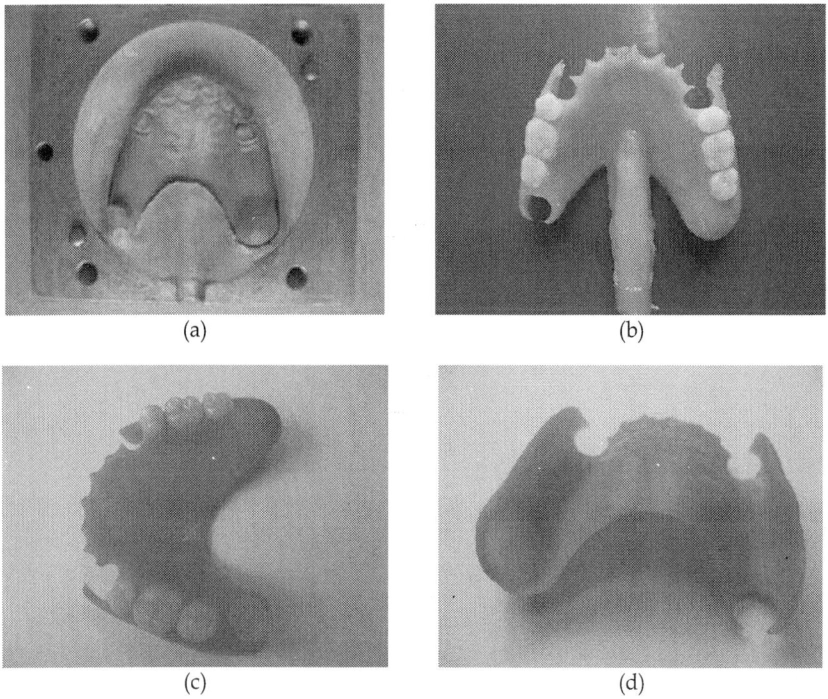

Fig. 20. Removable partial dentures made of a super-flexible polyamide (a) The model with retentive tuberosity embedded in the flask, (b) The denture immediately after unwrapping, (c), (d) The flexible removable partial denture

When manufacturing polyamidic dentures, the support elements blend in with the rest of the denture, as they are made of the same material (Szalina, 2005). - Fig. 21, 22.

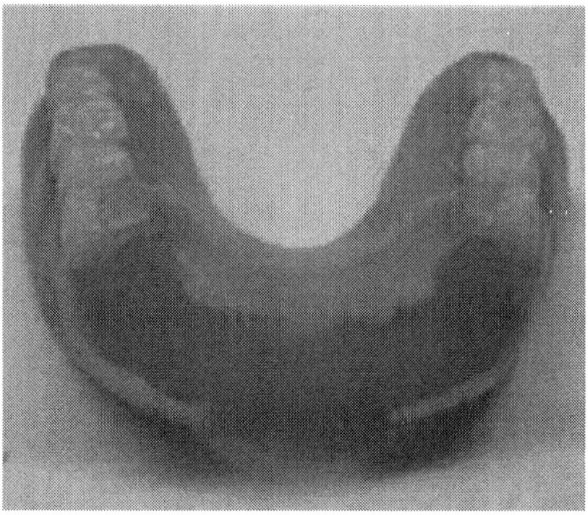

Fig. 21. Medium-low flexibility thermoplastic polyamide denture

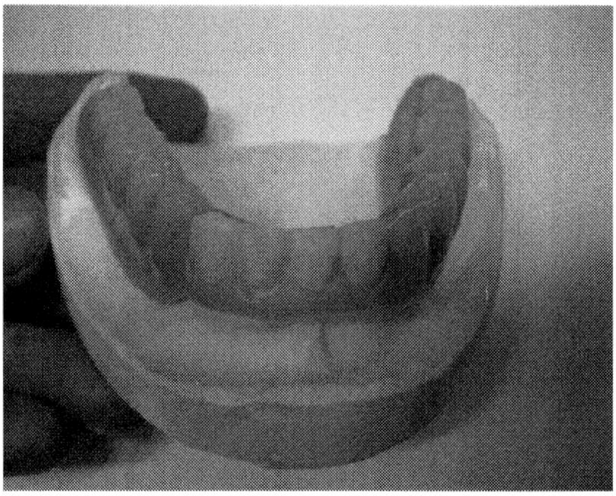

Fig. 22. Another medium-low flexibility thermoplastic polyamide denture

The superflexible polyamide resin is extremely elastic, virtually unbreakable, monomer-free, lightweight and impervious to oral fluids (Fig. 23, 24). The medium-low-flexibility polyamide is a half-soft material which has much wider range of use, being the ultimate cast thermoplastic for removable partials and offering the patient superior comfort, esthetics with no metallic taste (Fig. 21, 22). It is easy to polish and adjust, it can be added to or relined in office or laboratory. As previously shown, in certain cases we used pre-formed clasps, made of nylon, its composition being similar to that of the polyamidic resin used for denture manufacturing, which is adapted to the tooth by heating. It can be used for classical dentures, with metal framework, or it can be associated with injected thermoplastic resins. In other cases we chose to make the clasps of the same thermoplastic resin as the saddles of from acetal resin.

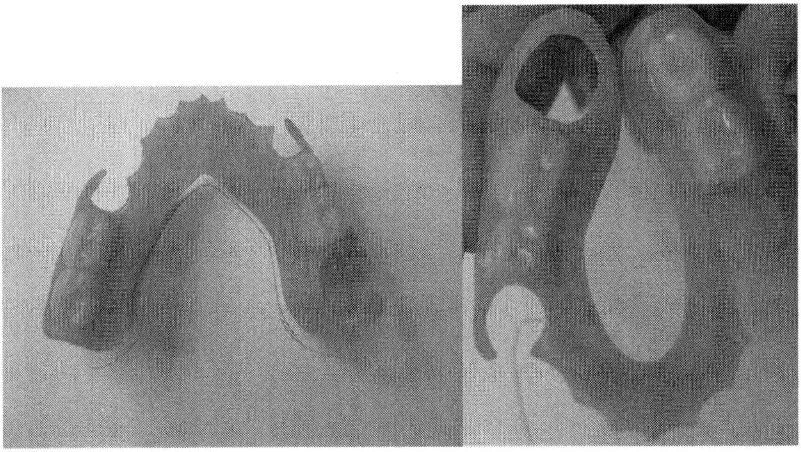

Fig. 23. Superflexible polyamidic partial dentures

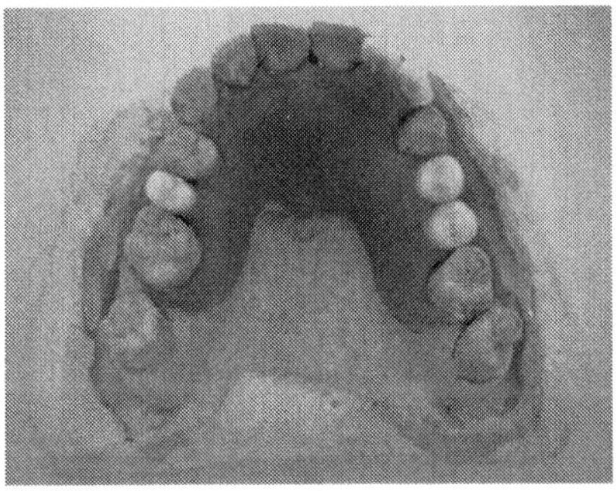

(a)

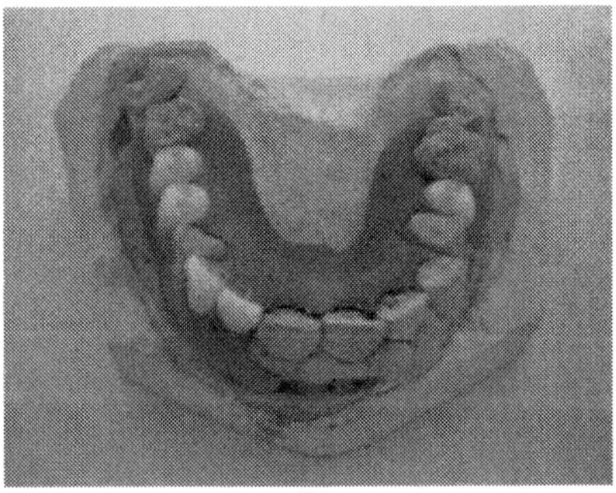

(b)

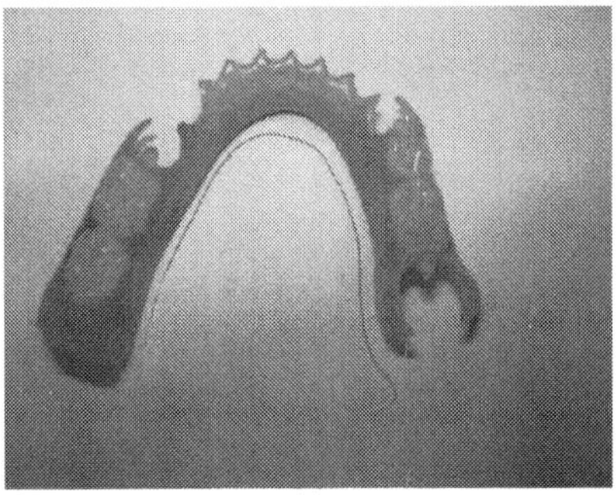

(c)

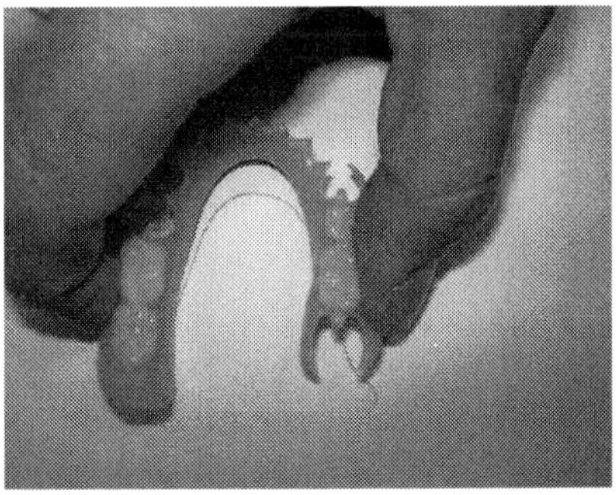

(d)

Fig. 24. (a), (b). Maxillary patterns, (c). The super-flexible polyamide denture , (d). The final flexibility test

7. Errors in manufacturing thermoplastic resins dentures

Errors might occur when manufacturing thermoplastic resins dentures: the insufficient pressure at injection, which leads to lack of substance, poor polishing or too thick saddles being some of the causes (Fig. 25).

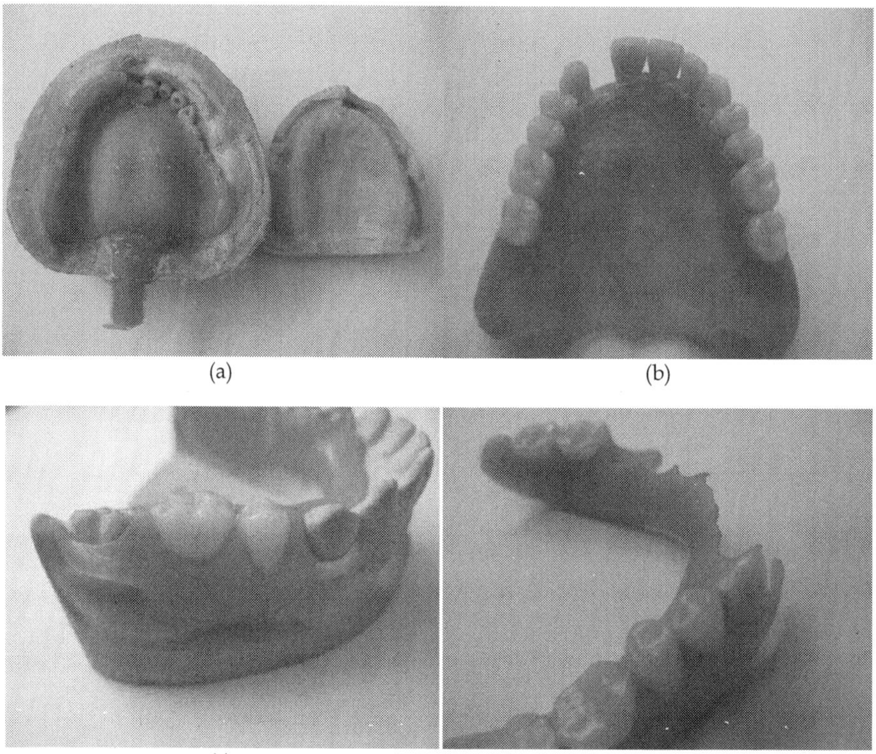

(a) (b) (c) (d)

Fig. 25. Errors which might occur when manufacturing dentures from thermoplastic resins; (a), (b) Lack of substance, (c), (d) Poor polishing

8. Myofunctional therapy devices

Recently much attention has been paid to the problem of controlling dento-facial growth interferences. The negative effects of mouth breathing, abnormal lip and tongue function and incorrect swallowing patterns on cranio-facial development in the mixed dentition period is well known. Correcting these myofunctional habits has been shown to improve cranio-facial growth and lessen the severity of malocclusion.

The purpose of myofunctional therapy is to retrain the muscles of swallowing, synchronize the movements of the swallow, and to obtain a normal resting posture of the tongue, lips, and jaw. Treatment may be received before, during or following orthodontic treatment (Quadrelli et al, 2002). The age range can be from 4 through 50 years of age, with the most typical age between 8 and 16 years.

A lot of appliances were manufactured in order to treat this problem, the main objective of all these myofunctional appliances being to eliminate oral dysfunction and to establish muscular balance. There is a definite place for these appliances in orthodontics today because they are simple and economical, but the cases need to be carefully selected, and the operator needs to be well trained in their use.

Some products do not require manufacture in the laboratory and are made in a universal size for all children 6-11 years of age (mixed dentition stage), allowing orthodontic treatment to be implemented earlier and at lower cost.

Some of these are flexible appliances are made of a thermoplastic silicone polycarbonate-urethane which is a ground-breaking copolymer that combines the biocompatibility and biostability of conventional silicone elastomers with the processability and toughness of thermoplastic polycarbonate-urethanes.

The silicone soft segment works synergistically with polycarbonate-based polyurethanes to improve in vitro and in vivo stability. It's strength is comparable to traditional polycarbonate urethanes and the biostability is due to the silicone soft segment and end groups.

It's adaptable to various fabrication techniques to accommodate many different device shapes and capable of being extruded and injection or compression molded, as well as solvent bonded, dipped, coated and sprayed. Additional surface processing steps after the device component fabrication is not needed.

9. Conclusion

Solving partial edentations with metal-free removable partial dentures represents a modern alternative solution to classical metal framework dentures, having the advantage of being lightweight, flexible and much more comfortable for the patient (Wostmann et al., 2005). Metal-free removable partial dentures made of thermoplastic materials are biocompatible, nonirritant, sure, nontoxic, biologically inert, with superior esthetics, which make them rapidly integrate in dento-maxillary structure. They offer quality static and dynamic stability.

The effectiveness of the technique is given by the use of the same material in making the clasps or the use of ready-made clasps from the same material (Ardelean et al., 2007). Where the mechanical resistance of the structure comes first, the choice is an acetal resin for making the frame. Superflexible polyamide resin is especially indicated for retentive dental fields, which would normally create problems with the insertion and disinsertion of the removable partial dentures. Of the thermoplastic materials used by us for manufacturing removable partial dentures, using acetal resin flexible thermoplastic frame, is the most laborious, requiring most working steps, due to the fact that first step involves manufacturing the acetal frame, afterwards the acrylic saddles and artificial teeth being manufactured (Ardelean et al., 2010). A removable partial denture with the framework made of acetal resin should be quickly integrated into the dento-maxillary system and accepted by the patient due to its reduced volume, esthetic and flexible clasps. Such a removable denture is a comfortable solution for the partial edentulous patient, achieving the principles of static and dynamic maintenance and stability. The partial dentures made of acetal resin thermoplastic materials are not bulky, resin frameworks may be as thin as 0.3-0.5 mm, clasps are flexible and esthetic, being rapidly integrated in the DMA structure, thus representing the most comfortable solution for the patient.

In small dentures, it provides excellent support, static and dynamic stability. The material is opaque, thus avoiding the translucency of dark backgrounds and making possible the manufacturing of matching bases only 3 mm thick, being recommended for injected clasp partial dentures, sliding or telescopic dentures, lingual splints or sport mouthguards.

A particular advantage of a removable partial denture made of acetal resin applies to patients with large oral defects as a result of a maxillectomy procedure, who are due to have postoperative radiotherapy and need to have the density of the defect restored to ensure standardized radiation distribution. Different types of boluses may be used for restoration but a stent is usually needed as a support. Traditional metal-clasp retained stents are discarded in such cases as the clasps cause backscatter of the radiation beams. Acetal resin is a radiolucent material suitable for making a stent with clasps or even a RPD to retain the bolus.

In the case of Kemeny-type acetalic dentures, the effectiveness of the technology is given by making artificial teeth of the same material, in the same step as the rest of the denture. As the material is not translucent, it is mainly suitable for dealing with lateral edentations but it can be used temporarily, in the frontal area as well, in those clinical cases where short-term esthetic aspect is not important.

Unlike conventional acrylates, thermoplastic resins have several advantages: long-term performance, stability, resistance to deformation, resistance to wear, excellent tolerance, resistance to solvents, absence or low quantity of allergy-inducing residual monomer, lack of porosity, thus preventing the development of microorganisms and deposits, all of which, together with maintaining size and colour in time are very important characteristics for removable dentures, presenting a high degree of flexibility and resistance, permitting the addition of elastomers for increased elasticity or reinforcement with fiberglass, in order to increase their physical splinterty quality; some of them can also be repaired or rebased (Szalina, 2006).

The advantages of using the molding-injection system lay in the fact that the resin is delivered in a cartridge, thus excluding mixture errors with long-term shape stability, reduces contraction, and gives mechanical resistance to ageing (Parvizi et al., 2004).

Thermoplastic resins have been used in dental medicine for fifty years. In the meantime, their use has spread due to their superior characteristics. Their ongoing development has yielded new classes of more and more advanced materials and technologies, which make possible the manufacturing of dentures with better splinting properties than traditional dentures.

Processing technology is based on the thermal plasticization of the material, in the absence of any chemical reaction. The possibility of injection-molding of the plastified material has opened new perspectives in the technology of total and partial removable dentures. The technology of injection-molding hasn't been widely used in dental technique labs yet, as it requires special injection-molding devices.

As this class of materials, as well as the processing devices, have been continuously perfected, their future applicability in dental medicine will keep spreading.

Most probably, further chemical development of elastomeric and polymeric materials will enlarge the domain of clinical applications of thermoplastics in dentistry.

10. References

Arikan, A.; Ozkan, Y.K.; Arda, T. & Akalin, B. (2005). An in vitro investigation of water sorption and solubility of two acetal denture base materials. *European Journal of Prosthodontics and Restorative Dentistry*, Vol. 13, No. 3, pp. 119-122, (September 2005), ISSN 0965-7452.

Ardelean, L.; Bortun, C. & Motoc, M. (2007). Metal-free removable partial dentures made of a thermoplastic acetal resin and two polyamide resins. *Materiale Plastice*, Vol. 44, No. 4, pp. 345-348, (December 2007), ISSN 0025-5289.

Ardelean, L; Bortun C.; Motoc M. & Rusu L. (2010). Alternative technologies for dentures manufacturing using different types of resins. *Materiale Plastice*, Vol. 47, No. 4, pp. 433-435, (December 2010), ISSN 0025-5289.

Bortun, C.; Lakatos, S.; Sandu L.; Negrutiu, M. & Ardelean, L. (2006). Metal-free removable partial dentures made of thermoplastic materials. *Timisoara Medical Journal*, Vol. 56, No. 1, pp. 80-88, (January 2006), ISSN 0493-3079.

Donovan, T. & Cho G.C. (2003). Esthetic considerations with removable partial dentures. *Journal of the California Dental Association*, Vol. 31, No.7, pp. 551-557 (July 2003), ISSN 1043-2256.

Quadrelli, C.; Gheorghiu, M., Marchetti C. & Ghiglione V. (2002). Early myofunctional approach to skeletal Class II. *Mondo Ortodontico*, Vol. 27, No. 2, pp. 109-122 (April 2002), ISSN 0319-2000.

Negrutiu, M.; Bratu, D. & Rominu, M. (2001). Polymers used in technology of removable dentures, *Romanian Journal of Stomatology*, Vol. 4, No.1, pp. 30-41, (March 2001), ISSN 1843-0805.

Negrutiu, M.; Sinescu, C.; Rominu, M.; Pop, D.; Lakatos, S. (2005). Thermoplastic resins for flexible framework removable partial dentures. *Timisoara Medical Journal*, Vol. 55. No. 3, pp. 295-299, (September 2005), ISSN 0493-3079.

Ozkan, Y.; Arikan, A.; Akalin, B. & Arda T. (2005). A study to assess the colour stability of acetal resins subjected to thermocycling, *European Journal of Prosthodontics and Restorative Dentistry*, Vol. 13, No. 1, pp. 10-14, (March 2005), ISSN 0965-7452.

Parvizi, A.; Lindquist, T.; Schneider, R.; Williamson, D.; Boyer, D. & Dawson, D.V. (2004). Comparison of the dimensional accurancy of injection-molded denture base materials to that of conventional pressure-pack acrylic resin. *Journal of Prosthodontics*, Vol. 13, No. 2, pp. 83-89, (June 2004), ISSN 1532-849X.

Phoenix, R.D.; Mansueto, M.A.; Ackerman, N,A, & Jones, R.E. (2004). Evaluation of mechanical and thermal properties of commonly used denture base resins, *Journal of Prosthodontics*, Vol. 13, No. 1, pp. 17-24, (March 2004), ISSN 1532-849X.

Szalina, L.A. (2005). Tehnologia executarii protezelor termoplastice Flexite. *Dentis*, Vol. 4, No. 3-4, pp. 36, (December 2005), ISSN 1583-0896.

Szalina, L.A. (2006). Posibilitati de reparatie a protezelor termoplastice Flexite. *Dentis*, Vol. 5, No. 1, pp. 38, (March 2006), ISSN 1583-0896.

Wostmann, B.; Budtz-Jorgensen, E.; Jepson, N.; Mushimoto, E.; Palmqvist, S.; Sofou, A. & Owall, B. (2005). Indications for removable partial dentures: a literature review. *International Journal of Prosthodontics*, Vol. 18, No. 2, pp. 139-145, (June 2005), ISSN 0893-2174.

4

High Performance Thermoplastic/Thermosetting Composites Microstructure and Processing Design Based on Phase Separation

Yuanze Xu* and Xiujuan Zhang

College of Chem. & Chem. Eng. Xiamen University, Xiamen, Dept. Macromol. Sci. Fudan University, Shanghai, China

1. Introduction

The merging of two research fields of thermoplastics and thermosets (TP/TS) is creating high performance composites with high rigidity, toughness and thermal stability etc. for versatile applications (Martuscelli,1996). However, most TP/TS mixtures do not always show the synergistic effects of rigidity and toughness as one expected (Bulknall, 1985; Douglas et al, 1991; Chen, 1995; Hayes et al, 200). On the other hand, the nature sets very elegant examples of strong solid structure, e.g. animal bones and wood plants, where sophisticated inhomogeneous structures are introduced to disperse the internal stress, yet keeping some interfacial strength to eliminating the micro-cracking or fracture propagation to achieve both high rigidity and toughness (Maximilien et al, 2010). Figure 1a is an example of SEM picture of animal bone, shown in Figure 1b is an illustrative example of a calculated bi-continuous phase separated structure of TP modified TS composites which resembles the nature occurring fine material structures.

This inspires us to improve the structures of our TP/TS systems, which will phase separate in a controllable way thermodynamically. Even very compatible mixtures of TP in TS matrices of practical importance will become incompatible during curing at certain stage, which is a prerequisite of toughening the brittle TS matrix. This process is so called cure reaction induced phase separation (CIPS) (Inoue 1995). CIPS provides an ingenious approach to realize controllable multi-scale phase morphology from nano-, micro to macro-scales (Hedrick et al, 1985; Ritzenthaler, et al, 2002a; 2003b), especially, the bi-continuous and/or phase inverted morphologies generated via spinodal decomposition may create some favorable structures which are critical in toughing the TS matrices (Cho,1993;Girard-Reydet et al, 1997; Oyanguren et al, 1999).

The separated phase morphologies will freeze through the matrix gelation or vitrification near Tg which become the final structures in TP modified TS composite materials. In this sense, our major task is to find some general relations between the miscibility, phase separated morphology evolvement, on one side, and the chemical structures of components and cure

* Corresponding Author

processing parameters in time-temperature windows on the other side. This is still a challenging goal up to now because of our limited knowledge of phase separation and the gaps between the CIPS research status and processing practices. To achieve the goal of morphology/process design, the combined efforts of multi-scientific approaches are employed both experimentally and theoretically on many practical systems of TP/TS/hardener. The results are promising as reflected in some of our recent publications (Zhang,2008;Zhang et al, 2008a;2007b;2008c;2006d;2006e) as well as in the new composite materials achieved based on this research for the structural parts of aircraft (Yi,2006;Yi et al, 2008;Yi& An,2008;Yi,2009).

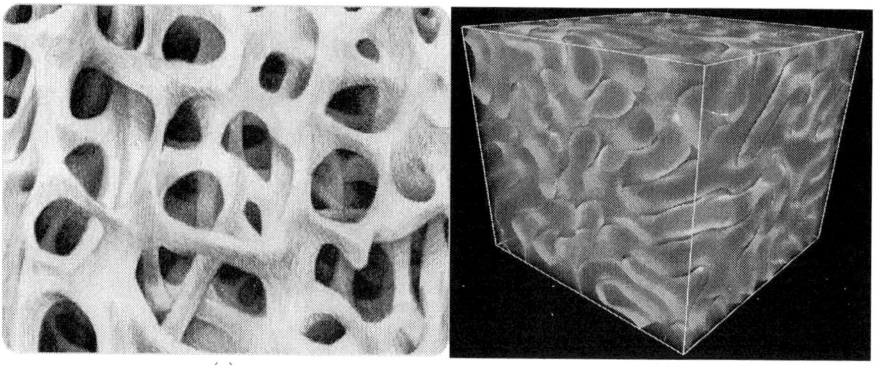

(a) (b)

Fig. 1. Skeletal structure of animal bone and simulated bi-continuous phase structure of some composite material. (a) Bone structure. (b) Bi-continuous phase of a composites in phase separation

The present chapter will provide a concise summary of our up-to-date original contributions with relevant literatures in this field emphasizing the breakthrough in approaches to understand and control the CIPS during processing.

2. Multi-scale design of the ideal composites

In macroscopic scale various reinforced TP or TS composites are designed with various graphite, glass and/other fibrous texture and alignments, where the matrices are treated as homogeneous. It is understood more and more that the rigidity and toughness of matrices are also of practical importance. The high rigidity & toughness require multi-scale controllable inhomogeneity of composites including bi-continuous and/or phase inverted micro-network with adjustable interface (Girard-Reydet et al, 1997; Oyanguren et al, 1999). After the concept of CIPS was proposed in 80s of last century (Inoue 1995), many experimental and theoretical investigations have been carried out to study the CIPS process in mixtures of tightly crosslinked TS matrices with ductile TP which is now assumed a promising alternative to rubber toughening approach, particularly when high value of elastic modulus, strength and glass transition temperature Tg are required.

For the continuous and/or inverse minor TP phase generated during CIPS and fixed in the thermosetting matrix, the most efficient toughening candidates are those that are able to plastically deform and can withstand the crack bridging/pinning or crack deflection and yielding (Pearson & Yee, 1993). Proper adhesion between the interfaces of TP and TS phases is very crucial for effective toughening of the TS matrix. Poor adhesion will render the TP phase premature de-bonding before expected deformations occur, while over strong adhesion between TS and TP interface will restrain the extent of ductile deformation undergone by TP phase because of the excessive constraint brought by TS phase. This constraint would limit the amount of material which could be involved in the bridging and deformation process (Williams, 1997; Girard-Reydet et al 1997; Oyanguren et al, 1999). Therefore the modifier TP must be chose in such a way that optimal affinity between TP and TS components is ascertained properly whereby CIPS can occur freely while cracking and/or delaminating can be restraint. Good combination of TP/TS with proper miscibility and cure cycle can display synergistic toughening effect. Since the mechanical properties of the toughened materials are closely related to their final morphologies, many studies have been focused on the qualitative or pseudo-quantitative thermodynamic and kinetic analysis of the phase separation mechanism (Girard et al, 1998; Gan et al 2003;Zucchi et al,2004) and morphological microstructure control as we will describe in next sections. At the same time, lots of the researches on the CIPS are limited to academic concern, the control of the morphology during cure cycle of commercial application targeting predefined mechanical properties is to be disclosed.

The well studied process is to mix the high performance TP into the matrix to form a homogeneous solution of the two components, and the subsequent cure generates the phase separated morphological structures. This process is so called *in situ* approach, wherein the phase separated structures are distributed evenly in the matrix. Drawbacks of the *in situ* approach are the substantial increase of the matrix viscosity, whereby the handling and processing window of the composites are significantly deteriorated (Cheng et al, 2009) and, also, the adhesion and modulus match of TP/TS to strength reinforcer fiber may be not as good as TS alone. Contrary to the *in situ* toughing, *ex situ* toughing is an innovative concept of a spatially localized toughening concept (Yi et al), whereby the composites are fabricated by interleaving TP containing layers and TS impreged carbon fiber layers initially. During the CIPS and the interlaminar diffusion, formed a spatial gradient of toughness and rigidity between highly toughened layers to the non-toughened, high rigidity graphite plies. Thus, the basic idea of *ex situ* concept sets a good example of the multi-scale design by controlled processing. As a part of the *ex situ* manufacturing of composite, our work is focused on the CIPS morphology evolvement, chemorheology, the time-temperature dependence of CIPS process in broad time-temperature processing window with a combined approach of experiment and theory.

3. TP/TS miscibility & cure induced phase separation (CIPS)

3.1 Strategies and approaches

The curing process involves several transformation, e.g. from the viscous clear liquid to a phase separated opaque dope solution and then to three dimensional chemical gels, finally to a vitrified solid. Phase separation occurs due to the change of the mixing enthalpy and entropy which arise from the size increase of the thermosetting oligomers. Upon gelation, the thermosetting systems lose fluidity permanently, and the phase domain loci are fixed, while the domains sizes keep growing through the component inter-diffusion. The gel times predefines the flowing operation window, e.g. the pot life, all the liquid molding methods such as RTM (resin transfer molding) have to be finished before permanent gelation. Vitrification is solidification that is defined to occur when Tg reaches cure temperature. Upon vitrification, the mobility of the composite molecules is restraint seriously, further polymerizations are restrained. So for TS composites processing, all the cure temperature usually are set above the material Tg for efficiency concerns. And the ultimate vitrification temperature set the material application temperature, once the environment temperature exceeds the material ultimate vitrification temperature, the material will become elastic rubber and the modulus and strength losses are serious, e.g. material yielding. To get material with desired morphology and properties, it is necessary to exploit these transformations to design the cure time-temperature processing routine so as to get the wanted ultimate structures and properties. In the following sections we describe the experimental approaches to detect and quantify the key parameters, from morphology to chemorheology in wide time/temperature space as presented by TTT (time–temperature-transformation) diagram (Grillet et al,1992; Simon&Gillham,1994). Our breakthroughs rely on the systematic studies on the phase separation time-temperature dependences during curing with the variations of the material parameters in a broad time/temperature space (Zhang et al, 2008a; 2007b; 2008c; 2006d; 2006e). To predict the possibility of various morphologies generated during cure, thermodynamic analysis based on Flory-Huggins-Staverman (F-H-S) theory was employed, which considers the free energy changing during demixing due to the enthalpy and entropy change during curing. It describes the spinodal and bimodal decomposition lines (Riccardi et al, 1994) while the relations to the structural parameters are to be clarified. Flory-Huggins-Staverman theory is widely used to describe the CIPS processes successfully in systems with UCST (up critical solution temperature) type phase behavior, like prediction the phase diagram together with the help of laboratory experiments of cloud point fitting (Riccardi et al, 1996; Girard et al, 1998; Ileana et al, 2004; Riccardi et al, 2004a; Reccardi et al, 2004b) and explain the morphology generated during cure, but further fundamental theoretical efforts are still needed. This is because at moment, the theoretical analysis is only successful for UCST systems, while those with LCST (lower critical solution temperature) type phase behavior are less well understood. The absence of property theoretical interpretation of LCST systems arises from the present theory approximation, wherein simplified assumption was made that the components are incompressible regarding to temperature, pressure and composition, while compressibility and special interaction place essential effect on the LCST phase diagram. The morphology evolution may be analyzed using time dependent Ginzberg- Landau equation (TDGL) of phase separation dynamics (Taniguchi&Onuki 1996) and the Viscoelastic model based on the fluid model(Tanaka 1997). Both models describe the morphological pattern evolvements, provides no clues yet about the temperature dependence of phase separation, though. All efforts towards processing condition will enable us better control CIPS and end-use properties of TP/TS composites.

In our present research, we focus on the curing systems which are widely studied and of great practical importance. Shown in Table 1 are chemicals we used throughout our study.

Abbreviations	Detailed chemical name	Chemical structure and characteristics
TGDDM	TGDDM: N, N, N', N'-tetraglycidyl-4, 4'-diaminodiphenylmethane, used as the matrix resin	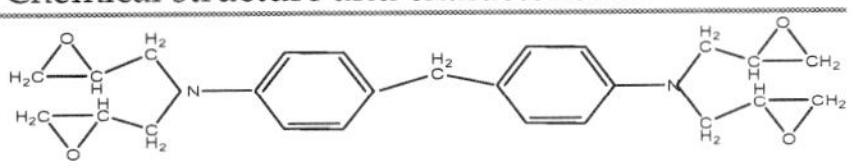
DGEBA	Diglycidyl ether of phenol A, used as the matrix resin	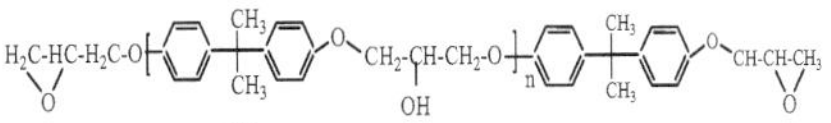
AroCyL10	Ethylidene di-4,1-phenylene ester, used as the matrix resin	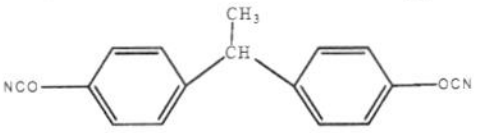
BMI	4,4′-bismaleimidodiphenylmethane, used as the matrix resin	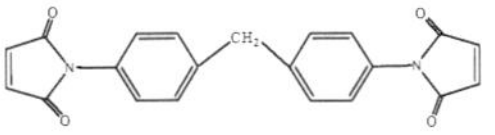
DBA	o,o′-diallyl bisphenol A, used as cure agent of BMI	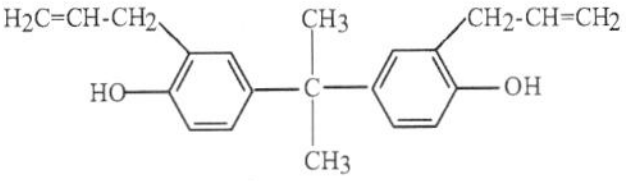
DDM	4,4′-diaminodiphenlyene methane, used as epoxy hardener	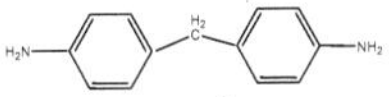
DDS	4,4'-diaminodiphenylsulfone, used as cure agent for epoxy resin	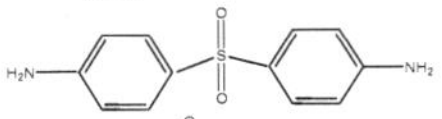
MTHPA	Methyl tetrahydrophthalic anhydride, used as epoxy hardener	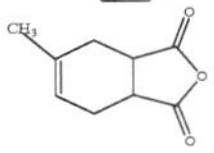
BDMA	Bipheylenemethylamine, initiator for the cure of DGEBA and MTHPA	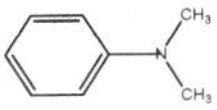
PEI	Poly(ether imide), used as modifier,	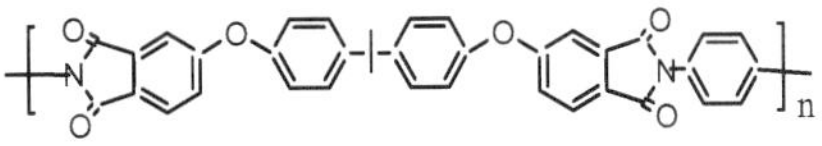
PEI1	Ploy(ether imide), used as modifier, provided by Prof. Shanjun Li	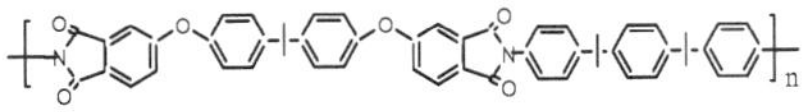
PES	Poly(ether sulphone) , used as modifier,	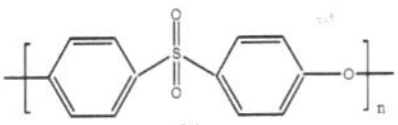
PSF	Polysulphone, used as modifier,	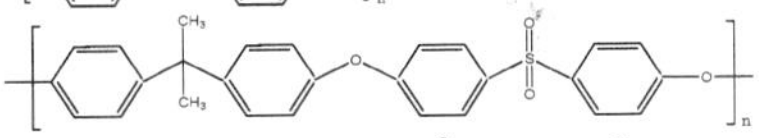
PES-C	Phenolphthalein poly(ether ether sulfone) , used as modifier,	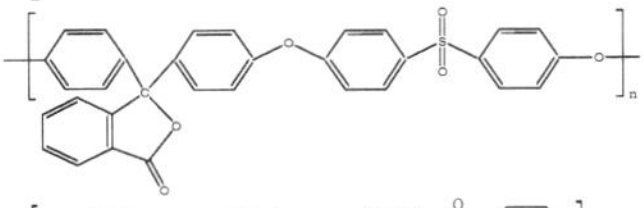
PEK-C	Phenolphthalein poly(ether ether ketone) , used as modifier,	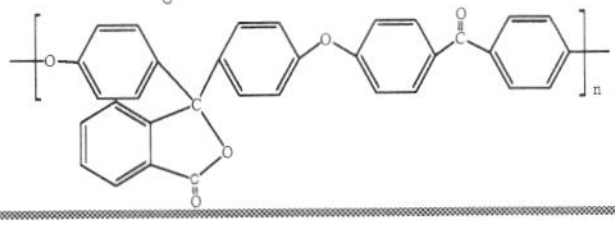

Table 1. Chemical structures of components employed in the present research

3.2 Detection of the cure induced phase separation process

The CIPS process can be observed in situ by different techniques, e.g. rheology (Xu et al, 2007; Zhang et al, 2006; Bonnet et al, 1999), small angle light scattering (SALS) and turbidity (Girard et al, 1998). Rheology can measure certain abrupt mechanical change upon phase separation, while SALS and turbidity trace the change of the optical mismatch between TP and TS rich domains during phase separation. SALS is most widely used to characterize the evolvement of domain size quantitatively in TP and TS blends (Bucknall&Gilbert, 1989; Girard et al, 1998; Gan et al 2003). In this work, we observed the early stage of CIPS in some TP/TS systems with a modified transmission optical microscope system(TOM) (Xu&Zhang,2007) and compared the microscopy method with the rheology and SALS approaches in the whole CIPS process. It was approved that the phase separation times can be determined by TOM. This enabled us to focus on the phase separation time/temperature dependence by considering the effects of TP molecular size and content, TP and TS structure, cure rate and cure mechanism and the stoichiometric ratio. The description of the whole cure time-temperature window of the thermoplastic modified thermosetting systems with the phase separation is of great importance in the composites processing industry.

3.3 Transmission optical approach

To observe the initial stage of phase separation, a lab-made computerized transmission optical microscope (TOM) system equipped with inversed optical design, long focusing objective lens and well controllable heating chamber was created which allows long term observation and data collection with a high resolution of 0.2μm in a wide working temperature range of RT~250°C (Xu&Zhang, 2007). The system can assign the onset of phase separation for systems with low refractive index difference and small domain size as low as 0.2μm, this won't be succeeded by usual TOM and SALS.

The samples for TOM observation were prepared by pressing the melt between two pieces of cover glass with a thickness of about 0.2mm. The moment when the morphological structure appeared was defined as the phase separation time t_{ps}. The values of t_{ps} at any particular temperature are the average of five measurements with observation errors of ±3% as measured in various TP modified TS systems (Zhang,2008; Zhang et al, 2008a, 2007b; 2008c; 2006d; 2006e).

We find that the phase separation times detected with different resolution optical lens are the same within a relative error of <±1.5% as shown in Table 2 for mixtures of DGEBA/DDM/PES and DGEBA/MTPHA/BDMA/PES with different PES content. The accuracy of the TOM approach to assign the onset of phase separation was also verified in various other systems like epoxy/DDS/PEK-C system, DGEBA/DDM/PEI, DGEBA/MTHPA/BDMA/PEK-C and (Cyanate ester)/TP systems. The fact that the t_{ps} values are independent on the magnification of the TOM proves that the onset of phase separation has been observed. If the initial phase domains were smaller than the resolution of the TOM system, one should see them earlier with higher magnification. The physical reason of limited initial domain size was explained as the nature of the cure induced spinodal decomposition by Inoue (Inoue, 1995; Ohnaga et al, 1994). Theoretically, the growing of TS size will drive the system into the unstable region in the phase diagram from the stable state, which will result in the decreasing of concentration fluctuation wave length and the rising quenching depth theoretically. But the successive increase of the quenching depth has not changed the regular concentration fluctuation sine wave length at intermediate cure rate, as observed in most of the CIPS processes with spinodal decomposition, where very regular and fine bi-continuous or inverse phase separation were observed, computer simulation also verified such phenomenon. Possibly the increasing quenching depth depresses the domain coarsening arising from interfacial tension and hydrodynamics in the early stage of CIPS. Experimentally, in the initial stage of phase separation, the domain size keeps constant because of the compromising effect of domain coarsening and quenching increasing. So it seems no size variation appears in the early stage of CIPS, the concentration fluctuation wave length kept constant, and the limited value of initial fluctuation wavelength becomes the periodic size of phase domains in the early stage of phase separation before domain growth driven by interfacial tension and hydrodynamic force so the early stage of the spinodal decomposition can be observed at similar t_{ps} using different TOM magnification.

Mixtures	T/℃	t_{ps}/s (1500X)	t_{ps}/s (1200X)	t_{ps}/s (760X)	Error
DGEBA/DDM/PES 10ppm	163	94	92	93	-1-1%
	153	141	142	143	-0.7-0.7%
	143	209	206	207	-0.5-1%
	133	319	318	320	-0.3-.3%
DGEBA/DDM/PES 15ppm	163	88	89	87	-1.1-1.1%
	153	143	140	139	-1.4-1.4%
	143	178	178	177	-0.2-0.2%
	133	265	267	263	-0.8-0.8%
DGEBA/MTHPA/BDMA/PES 10phr	163	94	92	93	-1-1%
	153	141	142	143	-0.7-0.7%
	143	209	206	207	-0.5-1%
	133	319	318	320	-0.3-0.3%
DGEBA/MTHPA/BDMA/PES 15phr	163	88	89	87	-1.1-1.1%
	153	143	140	139	-1.4-1.4%
	143	178	178	177	-0.2-0.2%
	133	265	267	263	-0.8-0.8%

Table 2. Illustrative show of the effect of TOM resolutions on the phase separation times in TP/TS/hardener mixtures

3.4 Dynamic rheological measurements

The rheological experiments were performed with a rotational rheometer with disposable parallel plates (gap 1 mm and diameter 40 mm) (ARES of TA Instrument Co.). The multiple frequency dynamic time sweeps were conducted under the isothermal conditions using the time resolved rheometric technique to collect automatically multi-frequency data as function of time(Mours&Winter,1994). The dynamic rheological data in Figure 2 were collected at frequencies 1, 2, 5, 10 and 20rad/s, respectively with an initial strain of 0.5 %. The strain level was automatically adjusted to maintain the torque response within the limit of the transducer.

Figure 2 shows the characteristic rheological profiles together with the corresponding morphological TOM micrographs of DGEBA/MTPHA/BDMA/PEK-C mixture along curing at 100°C. There are two critical transitions in the plots of *tan δ* at different frequencies in the left layout in Figure 2. The curves cross at the times of phase separation t_{ps} and chemical gelation t_{gel} respectively, wherein loss tangent becomes independent of frequency and the dynamic modulus shows power law function with frequency (Chambon et al,1986; Hess et al, 1988). This critical status corresponds to the gel point and the network shows fractal structure, as we have discussed more systematically in previous work (Zhang,2008; Zhang et al, 2008; Zhang&Xu, 2006). It was found that the critical gel at phase separation exhibits lower fractal dimension, indicating the looser structure of TP network as will be explained in Section 4.

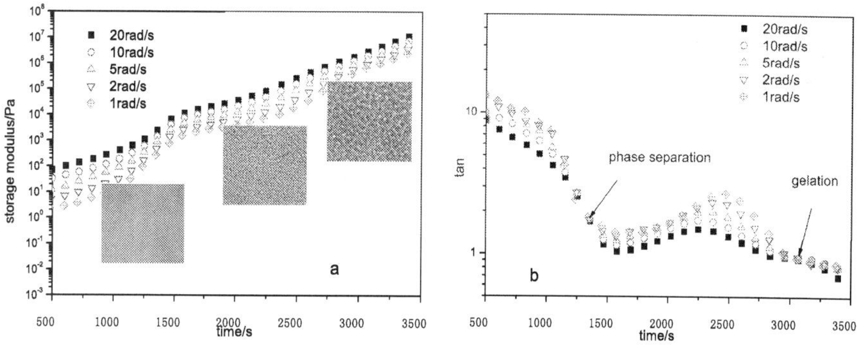

Fig. 2. Storage modulus G' (a) and tan δ (b) profiles and morphology evolvement of composite DGEBA/MTPHA/ BDMA/PEK-C 15phr systems cure isothermally at 100°C

3.5 Small angel light scattering

The CIPS of systems with enough optical mismatches were observed *in situ* by SALS (small angel light scattering) using HeNe laser light (λ = 632.8 nm). Samples were mounted in a temperature controllable hot-stage. The scattering pattern generated by the sample was visualized on screen and recorded by a CCD. The time upon which the scattering ring appears is defined as the phase separation time. SALS is a good approach to monitor the structures in micron scale in size, wherein the phase separated structures have big enough optical mismatch (Girard-Reydet et al, 1998; Yu et al, 2004). SALS approach is unable in detecting the beginning time of phase separation process in some systems we studies like TGDDM/DGEBA/DDSPEK-C systems, whereas high resolution TOP and rheology take more effective role in characterizing the morphology evolvement (Zhang, 2008; Zhang et al, 2008; Zhang & Xu, 2006). Actually, the SALS method observes heterogeneous structures with sizes above micron (Girard et al, 1998), while the present TOM approach has an optical resolution of 0. 2μm and can concurrently give the direct morphology evolvement information.

It is interesting to have the CIPS systems with enough optical mismatches where the combined detection approaches can be employed giving the comprehensive *in situ* information. As shown in Figure 3a of the DGEBA/MTHPA/BDMA/PES system, the phase separation time t_{ps} based on different observation means under various temperatures show a clear order: t_{ps} is earlier based on TOM, then rheology and SALS at last. In Figure 3b, DGEBA/MTHPA/BDMA/PEK-C shows similar sequential occurring between rheology and TOM approaches. It was observed that the phase separation time/temperature can be fit by the Arrhenius equation, whereby phase separation activation energy *Ea(ps)* was derived as shown in Table 3, which will be discussed in detail in the following sections. All the three approaches of SALS, rheology and TOM give slight different phase separation times, but the *Ea(ps)* in the two systems keeps almost constant, which was verified in lots of other TP modified TS systems. The discrepancy in the onset time of phase separation measured by different means probably comes from the fact that TOM detects smaller optical mismatch at initial stage of PS, while SALS needs sharper Interface and rheological measurement detects the overall mechanical response of the Components. Rheology is very very useful to get processing information.

Phase separation detecting means	PES 10phr		PEK-C10 phr		PEK-C15 phr	
	Ea(ps)/ kJ .mol-1	R	*Ea(ps)*/kJ .mol-1	R	*Ea(ps)*/ kJ .mol-1	R
TOM	77.8	0.997	77.6	0.999	77.1	0.999
Rheology	75.8	0.999	78.3	0.999	77.1	0.999
SALS	76.7	0.999	/	/	/	/

Table 3. *Ea (ps)* values of DGEBA/MTHPA/BDMA/TP systems detected by different means

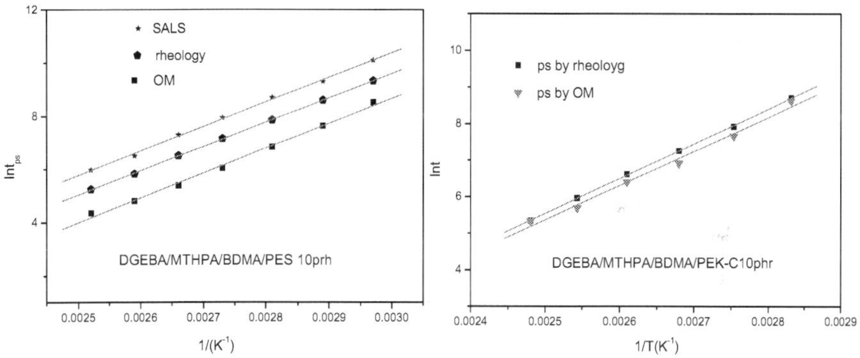

Fig. 3. Effect of detecting means on the time/temperature dependence for DGEBA/MTHPA/BDMA/TP systems

3.6 Differential scanning calorimetry

As a usual measurement method for TS curing study, DSC (differential scanning calorimetry) is a very convenient and effective means to monitor the thermosetting monomer cure kinetic, i.e. isothermal/dynamic cure rate, cure activation energy and cure enthalpy etc (Borrajo 1995; Bonnau, 2000). DSC is also widely used in the measurement of the polymeric material glass transition temperature Tg through the detection of heat capacity transition Cp of the material during temperature scanning. In our present work, DSC technique is used to test the Tg of TS matrix and TP modifier, measure the cure kinetic at different temperatures and the cure activation energy *Ea*.

Figure 4 shows the dynamic exothermic and isotheral cure kinetic curves of DGEBA/MTHPA/BDMA/PEK-C10 with 0.45% initiator BDMA. As shown in the left layout, the temperature scanning curves of systems with different scanning rates. It can be observed that the peak temperatures vary with the change of scanning rate. Displayed in Figure 4b is the isothermal cure kinetics of DGEBA/MTHAP/BDMA/PEK-C10phr at 100°CBy fitting such cure kinetic curve, information about cure reaction order and cure activation energy etc can be got.

By making the temperature scanning of the raw material with different scanning rate, like 5, 10, 15, 20, 25°C/min, then analyze the exothermic curve and recognize the peak temperatures T_p, the cure activation energy is calculated as below:

$$Ea = -R * \frac{d\ln(\beta / T_p^{\ 2})}{dT_p^{\ -1}} \quad (1)$$

Where β is the heating rate, *Tp* is the peak temperature, Ea is the cure activation energy and R is the gas constant.

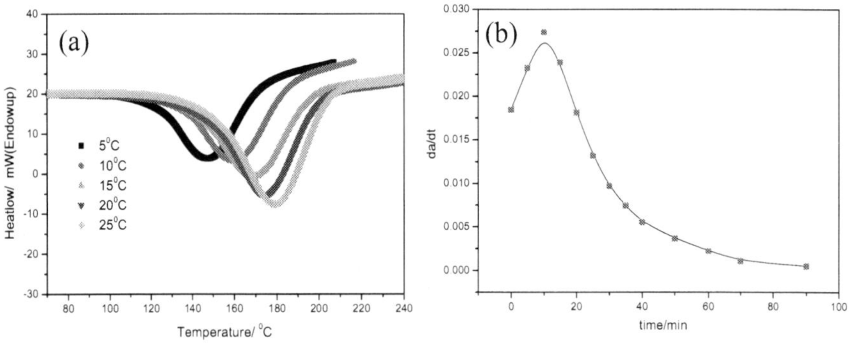

Fig. 4. DSC dynamic and isothermal cure kinetic monitor of DGEBA/MTHPA/BDMA/PEK-C10phr systems. (a) Exothermic curves. (b) Isothermal cure kinetics curve, r=1 at 100°C

4. CIPS of TP/TS in time-temperature window

4.1 Thermodynamic of CIPS process

Under the presence of TP, the cure reaction of thermosetting monomer with hardener is more complicated. According to the reaction mechanism, the polymerization of TS monomers falls into two groups: step polymerization and chain polymerization. Most TS monomers react via step polymerizations, which proceed by the stepwise reaction between the functional groups of reactive species wherein the size of the growing-TS molecules increases at a relatively slow pace. Thus, different-sized species present in the reaction system and the large macromolecules appears only at the later stage of conversion with broadening molecular weight distribution. Thermodynamically, phase separation takes place because of the increase of molecular size of the TS component and possible change of the interaction energy between the TP and TS species. With the growing of TS in size, the entropy of mixing decreases which is a very critical contribution to the free energy of mixing. Besides the entropy change, the mixing enthalpy generated from the interaction energy variation, which may decreases and/or increases, relying on the specific interaction between TS and TP components. When the mixing free energy rise above zero, the system is driven into unstable region and spinodal phase separation takes place, bi-continuous or phase inverted phases precipitate. Usually, the phase separation, gelation and vitrification appear sequentially, i.e., the phase separation takes place prior to the chemical gelation and, finally, vitrification as Tg increases with crosslinking going on.

With this physical scheme in mind, the thermodynamics of the cure induced phase separation in TP modified TS systems could be approximated in the framework of Flory-Huggins lattice theory (Riccardi et al, 1996; Girard et al, 1998; Ileana et al, 2004; Riccardi et al, 2004a; Riccardi et al, 2004b):

$$\Delta g = RT(\sum_i \frac{\phi_{TS,i}}{iV_{TS,i}} \ln \phi_{TS,i} + \frac{1}{V_{TP}} \sum_j \frac{\phi_{TP,j}}{j} \ln \phi_{TP,j}) + B\phi_{TS}\phi_{TP} \quad (2)$$

Where Δg is the mixing free energy in unit volume, R is gas constant, T is absolute temperature $\Phi_{TS,I}$, $\Phi_{TP,j}$ are the volume fractions of TS_i and TP_j respectively, B the interaction energy density. There are two parts in eq. 2 dominating the direction of CIPS process. In the first part, the entropy decreasing with growing of TS in size is always favorable to phase demixing. The second factor comes from mixing enthalpy by the exchange interaction between TP and TS, it may increase with cure of TS (favoring demixing), or keeps constant throughout cure, or decrease with polymerization (favoring mixing). The second part in Δg is very crucial for the pattern of phase diagram, whereby various polymer combinations manifest different diagram, like upper critical solution temperature (UCST), lower critical solution temperature (LCST) and even combined UCST/LCST phase diagram. It should be mentioned that the interaction energy coefficient χ in Flory-Huggins theory framework is not an effective characteristic measure in describing CIPS process, since χ is only correlated with temperature, not enough for the description of TP modified TS with continuous changing of TS density. Because of the limitation of the theoretical framework, up to know, most of the thermodynamic analysis of the polymer phase separation phenomena are qualitatively discussed with the help of χ.

In present work, we study the phase separation time-temperature dependence on the chemical environment, which may arise from the stoichiometric imbalance and component structures variation. We will use interaction energy density parameter B as shown in eq. 2 instead of χ to explain the interesting phenomena ever observed widely in the TP modified TS systems. Based on the Hildebrand-Scatchard-van Laar (HSL) theory, the interaction energy density was defined as below:

$$B = \frac{\chi RT}{V_{ref}} \quad (3)$$

Where V_{ref} is a random reference volume. The interaction energy density B contains two parts, the first part is the energy coming from energy exchange during mixing of different components, the second part arises from other energy change, like the extra energy that generated from the incompressibility, hydrogen bonding etc. There are several approaches to calculate or estimate the interaction energy density B, such as experimental testing the binary phase diagram, then fitting the mixing diagram with certain theoretical equation; test the mixing energy with analogous chemicals and the estimation based on solubility parameters. Experimental determinations of the interaction energy density parameter B are very burdensome, sometimes difficult to do. In our present discussion, we will estimate B value by the solubility parameter approach. The solubility approximation simplifies the complex interaction energy expression between different components to the internal pure component interaction and inter-component interaction, by assuming that inter-component interaction is a geometric mean of each pure components interaction:

$$\delta_{TP/TS}{}^2 = \delta_{TP}\delta_{TS} \tag{4}$$

Where δ is Hildebrand solubility parameter and is defined as: $\delta = \sqrt{\frac{\Delta E}{V_m}}$, ΔE is the molar energy of vaporization, V_m is the molar volume (Hildbrand, 1950). So it is possible to calculate the B value through the solubility parameters δ_{TP}, δ_{TS} by equation:

$$B = (\delta_{TP} - \delta_{TS})^2 \tag{5}$$

For small molecules it is practical to test δ value directly by measure the enthalpy change during vaporization which is extensively tabulated. Considering the fact that the heat of vaporization of a macromolecule can not be measured experimentally, it is a common practice to calculate δ using the corresponding value of the solubility parameter of its monomers as substitute. Fedor, Van Krevelen et al.'s group additivity, computer simulation (Lewin et al, 2010) and thermodynamic analysis (Fornasiero et al, 2002; Utracki & Simha, 2004) are also used widely in literature. Bicerano (Bicerano, 2002) developed a topological approach to calculate δ values based on the interaction index method, whereby group contribution is not necessary, δ can be deduced based on the polymeric chemical structure. We will illustrate the basic process to calculate δ as follow.

First, draw the skeletal structure of MTHPA in Figure 5a:

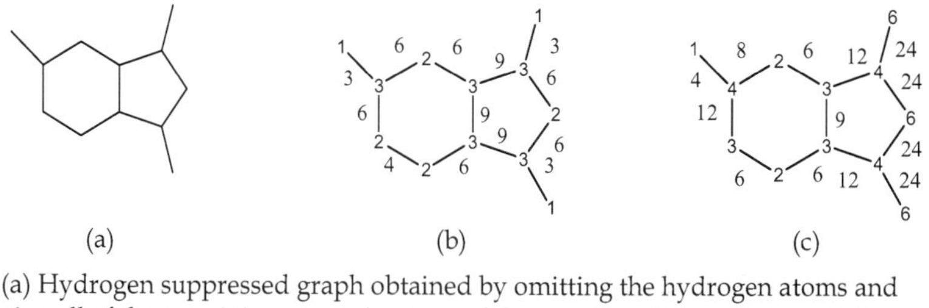

(a) (b) (c)

Fig. 5. (a) Hydrogen suppressed graph obtained by omitting the hydrogen atoms and connecting all of the remaining atoms (vertices of the graph) with bond(edges of the graph)

Then based on the simple atomic index γ of skeletal structure, non-hydrogen vertexes and the atom valence index γ^v to calculate the bond index $\beta_{ij} = \gamma_i * \gamma_j$ and valence index $\beta^v_{ij} = \gamma^v_i * \gamma^v_j$, the zero order connection index ${}^0\xi \equiv \sum_{vertices} (\frac{1}{\sqrt{\gamma}}) = \frac{3}{\sqrt{1}} + \frac{4}{\sqrt{2}} + \frac{5}{\sqrt{3}}$ =8.72 and ${}^0\xi^v \equiv \sum_{vertices} (\frac{1}{\sqrt{\gamma^v}}) = \frac{1}{\sqrt{1}} + \frac{2}{\sqrt{2}} + \frac{3}{\sqrt{3}} + \frac{3}{\sqrt{4}} + \frac{3}{\sqrt{6}}$ =6.87 and the first order connection index $\xi^1 \equiv \sum_{vertices} (\frac{1}{\sqrt{{}^1\beta}}) = \frac{3}{\sqrt{3}} + \frac{1}{\sqrt{4}} + \frac{6}{\sqrt{6}} + \frac{3}{\sqrt{9}}$ =5.68 and ${}^1\xi^v \equiv \sum_{vertices} (\frac{1}{\sqrt{{}^1\beta^v}})$ =4.09. Then put the connection index into equation for volume $V=3.64{*}^0\xi+9.799{*}^0\xi^v-8.54{*}^1\xi+21.69{*}^1\xi^v+0.98{*}N_{MV}$ (N_{MV} correction item) and cohesion energy $E_{Fedor}\approx 9882.5{*}^1\xi+358.7*(6*N_{atomic}+5*N_{group})$($N_{atom}$ and N_{group} correction): $\delta = \sqrt{\frac{E_{coh}}{V}}$ =21.8 $J^{1/2}/cm^{3/2}$. Then put δ into eq.4 to obtain B value. In our following discussion, all the interaction energy density parameters B are calculated based on Bicerano's approach.

4.2 Rheological characterization of the CIPS process in TP/TS mixture

TS/TP/harder is a typical asymmetric mixture. Most TP systems which are of practical importance usually have much higher molecular mass, higher modulus and Tg than that of TS monomers. The large difference in modulus/Tg and molar mass of TS and TP are responsible for the dynamical asymmetry in the relaxation and diffusion of chain which play a critical role in the early stage of phase separation process (Gan et al, 2003;Yu et al, 2004). At the beginning, TP was dissolved in the TS monomers homogenously. With cure going on, TS grows in size and the storage modulus built up higher as shown in Figure 2. When the miscibility between TP and TS jumped into the unstable region, TP precipitated out as inverse phase, or as nodular or dispersed phase depending on the TP volume fraction. The phase separation process is manifested by the abrupt increases of viscosity and modulus profiles. Then the viscosity and modulus evolve further upwards, which grow in a step way in the chemical gelation vicinity again. As can be found that phase separation occurs before chemical gelation - the infinite TS molecular structures originated from the percolation of the TS oligomers. As displayed in Figure 2b, two critical gel points were found, the first one occurs at the phase separation time t_{ps} which was verified by the morphological observation by TOM, the second critical gelation phenomenon appears at the chemical structure percolation point t_{gel} which was confirmed based on DSC isothermal exothermic process, solubility test and rheological study of the neat TS cure process (Macosko & Millerlb,1976).

It was observed in the beginning of cure, the TP/TS/harder blends have a storage modulus G' much lower than the loss modulus G'', which implies the viscous essentiality of the homogenous mixtures. Upon curing, G' grows in a more steep manner than G''. At the vicinity of phase separation point t_{ps}, log-log plots of G' and G'' versus ω became parallel and the loss angel factor tanδ became independent on frequency, implying a critical gel transition. TOM proved the appearance of the bi-continuous phase structure at this first transition. As the reaction proceeds, the shape of G' plots is further altered by the second rapid increase of G' which corresponds to the chemical gelation of TS oligomers confirmed by solubility tests and rheological monitoring on neat TS cure. After gelation, the margin of the domains became clearer gradually due to further phase separation and the coarsening of phase pattern can be observed depending on the interfacial tension and the viscosities of the phases.

4.3 The critical gel behaviors at t_{ps} and t_{gel}

As shown in Figure 6, at the vicinity of phase separation point t_{ps} and chemical gelation point t_{gel}, G' and G" involve in a parallel manner versus frequency ω in log-log plot, while loss angel factor tanδ become independent on frequency, implying the characteristic gel transitions, which Winter et al named as critical gel state (Scanlan & Winter, 1991). We will discuss such critical gel behaviors and its applications more systematically in following sections.

At both t_{ps} and t_{gel}, the dynamic scaling behaviors of storage modulus G' and the loss modulus G'' appears as shown in Figure 6, where G' and G'' versus frequency in the vicinity of t_{ps} and t_{gel} for system of DGEBA/MTHPA/BDMA/PEK-C 15phr isothermally curing at 100°C are displayed, the curves have been shifted horizontally by a factor A (see insert) for easier comparison. As can be seen G' and G'' show a power low relation through eq.6 ;

$$G' \propto G'' \propto \omega^{n} \tag{6}$$

Where ω is frequency; n is the critical relaxation index.

The difference between the critical gel at t_{ps} and t_{gel} was noticed that G' is less than G'' at t_{ps} while larger than G'' at t_{gel} as shown in Figure 6 and also reported in other systems (Scanlan&Winter, 1991). The second difference is in the slop of log G', G'' vs. log ω curves at critical states. The critical gel index at both of t_{ps} and t_{gel} are calculated and summarized in Table 4 for DGEBA/MTHPA/BDMA/PEK-C system with PEK-C of 15, 10, 5phr respectively. The scaling behaviors occur at both t_{ps} and t_{gel} and over the entire experimental temperature range of 90~130°C.

By analyzing the data in Table 4, it can be found that both of n_{ps} and n_{gel} are independent to cure temperatures within the experimental range, more significantly, n_{ps} is reasonably larger than n_{gel}. The existence of the power law index n value for a critical gel is a signature of its structure of self-similarity, for which the fractal geometry can be applied. Based on percolation theory, Muthukamar suggested that, the fractional dimension d_f can be represented in terms of power law index (Muthukumar, 1989):

$$d_f = \frac{(d+2)(d-2n)}{2(d-n)} \tag{7}$$

Where d_f is the fractal dimension and d is the space dimension.

Taking the values of n_{ps}, n_{gel} and space dimension of 3 to eq. 7, we obtained the value of fractional dimension at phase separation point $d_{f,ps}$ and chemical gelation point $d_{f,gel}$

respectively, both of which are summarized in Table 4. As was observed that bigger n gives a smaller fractional dimension implying looser crosslinking network. The values of $d_{f,gel}$ are within the range of most single TS gelation, as reported in the other networking systems (Eloundou,1996). The PS network is clearly looser than the chemical gelation ones. The TS crosslinking expels the less mobile high mass TP to form a sponge-like loose structure, where physical entanglement of TP macromolecules appears, causing storage modulus G′ jump or tangent loss drop. This may explain why the phase separation dimension fractional $d_{f,ps}$ values increase with the increase of TP content. It is also noticed that the chemical gelation factional dimension $d_{f,gel}$ also increases with the increase of TP content. The explanation may not be easy since multi-factors involved in the cure process with PS: component diffusion mobility, interface stability, hydrodynamic effect and the competition between phase separation dynamics and crosslinking mechanism (stepwise or chainwise polymerization) etc.

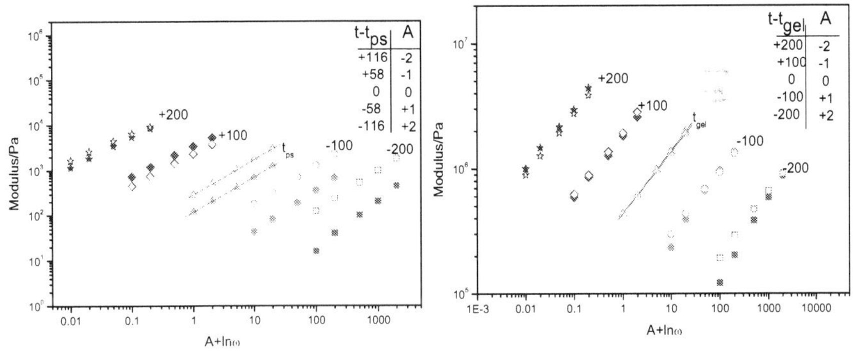

Fig. 6. Frequency response of G' (the solid symbols) and G'' (the open symbols) at the vicinity of t_{ps} and t_{gel} for DGEBA/MTHPA/BDMA/PEK-C 15 phr cure under 100°C

Temperature /°C	PEK-C 15phr				PEK-C 10phr				PEK-C 5phr	
	n_{ps}	$d_{f,psl}$	n_{gel}	$d_{f,gel}$	n_{ps}	$d_{f,psl}$	n_{gel}	$d_{f,gel}$	n_{gel}	$d_{f,gel}$
130	0.65	1.81	0.43	2.08	0.81	1.58	0.54	1.95	0.78	1.62
120	0.68	1.77	0.48	2.02	0.83	1.54	0.52	1.98	0.78	1.62
110	0.78	1.62	0.47	2.04	0.80	1.59	0.52	1.98	0.79	1.61
100	0.78	1.62	0.50	2.00	0.76	1.65	0.54	1.95	0.79	1.61
90	0.71	1.72	0.50	2.00	0.83	1.54	0.50	2.00	0.77	1.64

Table 4. Power law index at t_{ps} and t_{gel} under different temperatures for DGEBA/MTHPA/BDMA/PEK-C system

It is understood that the phase separation is incomplete after t_{ps} and some second component still left in the TP or TS rich phase for thermodynamic or kinetic reason. When the TP chain length is long enough like PEK-C we used in our study with molecular size far beyond the entanglement limit size, TP will behavior in such a way that the more the TP exists in TS rich phase, the denser the gel network will be. Similar scenarios can be imaged in the TP rich phase. Based on the kinetic and hydrodynamic approaches, we can explain the second phase separation observed usually in the CIPS process as was illustrative displayed in Figure 7, where TP and TS rich phase contain second phase respectively (Oyanguren et al, 1999). Here DGEBA reacts with hardener MTHPA at the presence of initiator BDMA through the chain polymerization, wherein large quantities of monomers exist upon phase separation, which will contribute greatly to the interface instability, second phase separations were observed in different temperature at different PEK-C contents as displayed in Figure 7a and Figure 7b. Low molecular weight TP show strong mobility after t_{ps}, also enhance the domain coarsening and second phase separation.

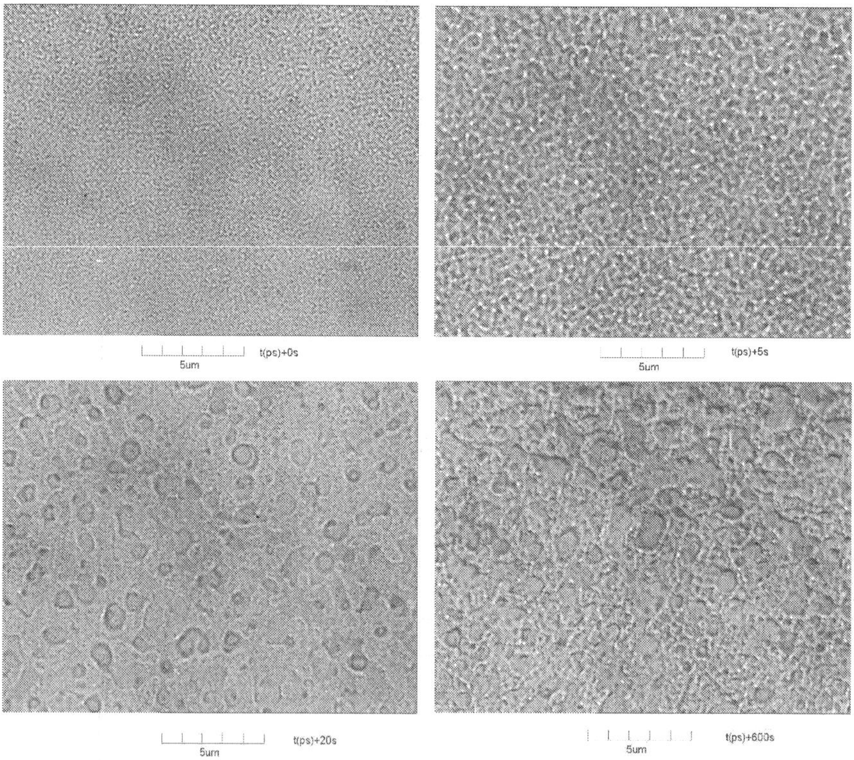

(a)

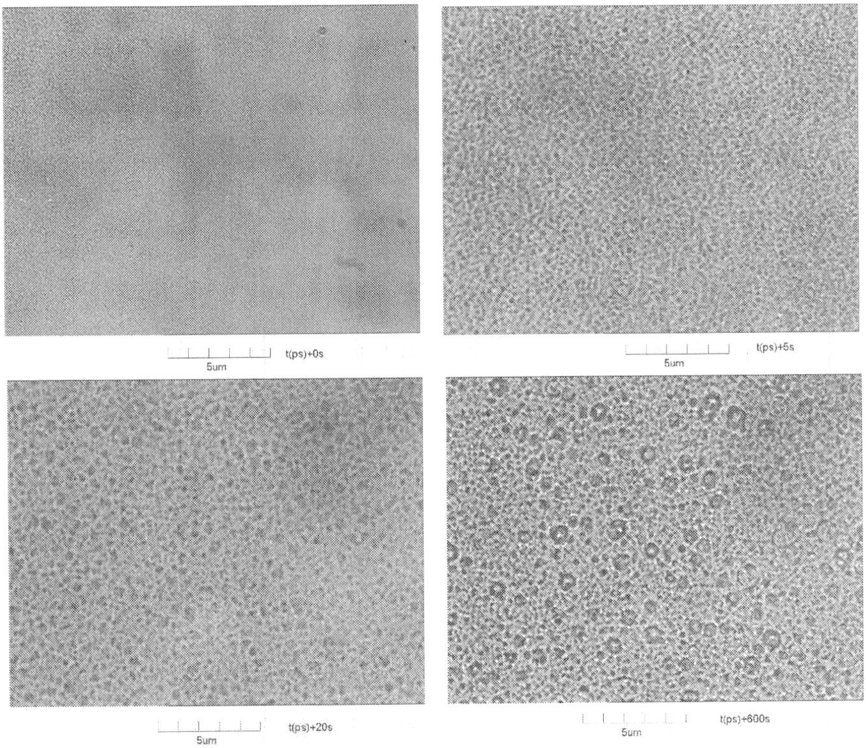

(b)

Fig. 7. Morphology evolvement of DGEBA/MTPHA/BDMA/PEK-C with different PEK-C content at 100°C (a)DGEBA/MTHPA/BDMA/PEK-C10prh
(b) DGEBA/MTHPA/BDMA/PEK-C 15prh

The critical phenomena during the CIPS process in other TP modified TS system were also observed, e.g. cyanate ester Arocy L-10/$Cu(Ac)_2$, DGEBA/DDM/PES and TGDDM/DGEBA/DDS/PEK-C systems etc.

5. Processing of TP/TS in time-temperature window, TTT diagram

5.1 Cure induced phase separation time-temperature dependence

Based on the determination of the phase separation times in different TP/TS/hardener systems at different temperature, we are able to study its temperature dependency. It was found that cure induced phase separation time & temperature relation fits into the Arrhenius form, irrespective the observation approaches of rheology, TOM or SALS:

$$\ln t_{ps} = \ln k + \frac{Ea(ps)}{RT} \tag{8}$$

Wherein t_{ps} is the phase separation time, *Ea(ps)* is the phase separation activation energy, *T* is the absolute temperature, and *R* is the universal gas constant.

The Arrhenius type dependence of CIPS time&temperature was observed in a various TS/TP/hardener mixtures, including the UCST (up critical solution temperature) and LCST(low critical solution temperature) types mixtures. The phase separation processes were monitored with TOM and rheology, some of them were also observed with SALS techniques. As were shown in Figure 3 and Table 3, the phase separation time-temperature dependency represented by phase separation activation energy *Ea(ps)* is independent on the detection means. The generality and factors which impair influence on *Ea(ps)* values have been discussed in detail in our published/unpublished works (Zhang,2008;Zhang et al, 2008a;2007b;2008c;2006d;2006e). It was observed that *Ea(ps)* values relay on the TP/TS internal chemical environment, like TS cure stoichiometric ratio, chemical structures of TP and TS, cure activation energy *Ea* etc, while the TP content, TP molecular size and cure rate show no obvious effect on the value of *Ea(ps)* in the TP/TS systems as far as TPs were employed as the minor part to toughen the TS matrix and cure reaction kinetics follows the Arrhenius type temperature dependence. Illustrative explanation on the time-temperature dependence of the CIPS in TP modified TS systems are made as following subsections.

5.2 Effect of TP contents and TP molecular size on *Ea (ps)*

Based on comprehensive study of the CIPS process of various TP/TS combinations in our research group, it was found that the phase separation energy *Ea(ps)* is independent on the TP content, wherein the TP is the minor modifier to toughen the TS matrix, systems with TS as minor component used as plasticizer are out of the scope of our research.

Illustratively Shown in Figure 8 are the drawings of $\ln t_{ps}$ vs. *1/T* for the LCST type system DGEBA/MTHPA/BDMA /PEK-C and UCST type system Cyanate ester AroCyL10/accelerator/PEI with different TP contents, respectively. The corresponding phase separation activation energy values of *Ea(ps)* derived from the slope of the $\ln t_{ps}$ vs. *1/T* plots are summarized in Table 5. As depicted in Figure 8, although both of the two systems contain different TP contents, phase separation times t_{ps} can be well correlated with cure temperature in Arrhenius type in wide cure reaction temperature. The *Ea(ps)* values shown in Table 5 are not sensitive to the TP content within the experimental range. The independency of *Ea(ps)* on the TP level can be understood from the physical origin of the CIPS process. Although the starting TP levels are different for the studied systems, but the driving force, and the chemical kinematics are same, which is manifested by the similar activation energy *Ea*.

Similar trends were also observed in other systems, e.g. Epoxy/anhydride/Initiator/PES (PEI) systems, Epoxy/amine /PES(PEI) systems, Epoxy/amine/PEK-C systems, Cyanate ester (AroCyL-10)/catalyst/PES system and Bismaleimide /TP systems. In all these systems, the phase separation times may vary because of the phase separation detection means, but the Arrhenius type phase separation time/temperature dependency Keeps unchanged. The exceptions could be mentioned that at high enough TP contents, the chemical reaction kinetics will be restraint because the high viscous environment will hinder the molecules free diffusion, under such event, Arrhenius type cure kinetic is not applicable, or when the cure temperature near glass temperature, diffusion incorporated model is necessary (Kim,2002).

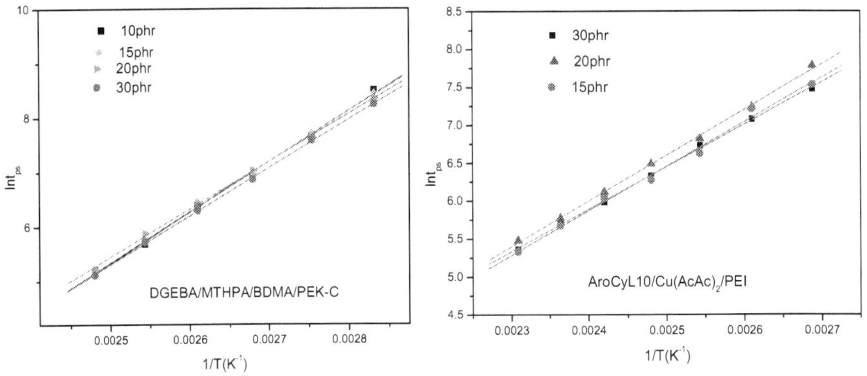

Fig. 8. Effect of TP content on the relation of $\ln t_{ps}$ *vs.* $1/T$ of DGEBA/MTHPA/BDMA/PEK-C and (AroCyL10) / Cu(AcAc)$_2$/PEI systems

TP content /phr	DGEBA/MTHPA/BDMA/PEK-C				(AroCyL10)/ Cu(AcAc)$_2$/PEI			
	Ea(ps, TOM)/ kJ .mol^{-1}	R	*Ea*/kJ .mol^{-1}	R	*Ea(ps, TOM)*/ kJ .mol^{-1}	R	*Ea*/kJ .mol^{-1}	R
5	77.1	0.999	73.3	0.999	n.d.	n.d.	60.1[a]	0.999
10	77.6	0.999	73.7	0.999	47.9	0.998	59.2	0.999
15	77.1	0.999	n.d.	n.d.	48.5	0.999	n.d.	n.d.
20	76.6	0.999	73.4	0.998	47.8	0.999	60.1	0.999

a: The reaction activity energy of both neat thermosetting monomer were determined by the gel point method

Table 5. *Ea(ps)* of DGEBA/MTHPA/BDMA/PEK-C and (AroCyL10)/Cu(AcAc)$_2$/PEI with different TP content

To understand the effect of the TP content on the component chain mobility and cure activation energy, the chemorhelogical processes of some TP modified TS systems were analyzed. The following chemorheological relationship (Halley & Mackay,1996) is applied to the TP/TS system with cure.

$$\ln \eta(t) = ln\eta_0 + kt \tag{9}$$

Wherein η_0 the initial viscosity of the system without curing , k reaction kinetic constant which is dependent to reaction temperature and cure reaction kinematic constant,η_0 and k follow Arrhenius form equation :

$$\eta_0 = \eta_\infty \exp\frac{E_\eta}{RT} \tag{10}$$

$$k = k_{\infty} \exp \frac{E_k}{RT} \quad (11)$$

The cure reaction process accompanied by the phase separation process of the TP modified TS systems with different TP content was monitored, the viscosity evolvement before phase separation was fitting by the Arrhenius chemorheological relation. Illustratively shown in Figure 9a are the complex viscosity η* profiles of DGEBA/MTHPA /PEK-C10phr with 0.9% initiator BDMA just before phase separation in the linear coordinate, displayed in Figure 9b is the same data but in the semi-log coordinator. The complete η* profiles under different cure temperature were displayed in the up right corner of Figure 9a. In Figure 9b of the semi-log coordinate, the η* curves were shifted on purpose by a factor of A, which is denoted by the number beside each curve. The straight forward trends of each curve in the semi-log coordinate imply that the cure process can well be fitted by the Arrhenius relationship. The cure activation energy E_k and viscous flow energy E_η were calculated and summarized in Table 6 respectively. It can be found in the present TP loading levels, E_k and E_η keep constant with the varying TP content. The results were confirmed in other characterization approaches, e.g. gelation and exothermic methods and in other systems. Flow viscosity itself is sensitive to the dragging effect exerting by TP macromolecules, while the viscous activation energy E_η value is related to smaller scale of TP chain segments. Considering the related mechanisms of E_η and *Ea (ps),* it is understandable that *Ea(ps)* keeps constant regardless the variation of TP content in low TP contents. The constant E_η values of the systems with different PEK-C loading level indicate that TP has no effect on the diffusion mobility of the TS oligomers during phase separation.

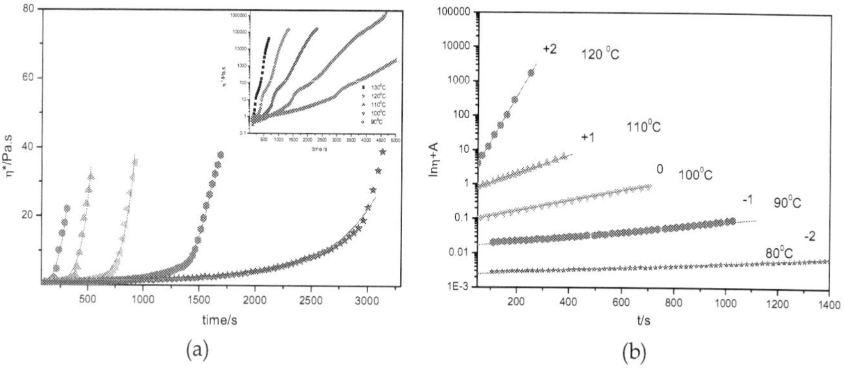

Fig. 9. Viscosity profiles of DGEBA/MTHPA/BDMA/PEK-C10ph system prior to phase separation under different cure temperatures. (a) η* profiles before t_{ps} in linear coordinator. (b) η* before t_{ps} in semi-log coordinator

PEK-C contents/phr	E_η/kJ .mol^{-1}	E_k/kJ .mol^{-1}	*Ea(gel, rheology)* / kJ .mol^{-1}	R
10	21.9	76.7	78.3	0.999
15	24.8	78.7	77.1	0.999
20	20.8	77.7	76.6	0.999

Table 6. E_η and E_k values of DGEBA/MTHPA/BDMA/PEK-C systems with different PEK-C content, BDMA content 0.9%

Shown in Table 7 are the phase separation time-temperature dependences of DGEBA/MTHPA/BDMA/PEK-C and DGEBA/MTHPA/ BDMA/PES systems with TP of difference molecular size, depicted by the intrinsic viscosity of [η]. It was observed that the $E_a(ps)$ are not altered obviously by the variation of TP molecular size. Theoretically, the miscibility of TP/TS changes with the size of TP as disclosed by Buckanall (Bucknall et al, 1994) in DGEBA/PES systems, which may change the startup of phase separation times t_{ps}, but the chemical reaction kinetics has not been altered significantly as shown in Table 7 denoted by cure activation energy *Ea,* wherein the effect of the variation of TP content resembles the effect of TP molecular size effect. In fact we found that some of the phase separation time/temperature data in published in literatures also followed the Arrhenius equation, e.g. from the data based on turbidity of castor oil modified epoxy system (Ruseckaite et al, 1993) and those based on light scattering of epoxy/dicyandiamide /PES systems (Kim et al, 1993).

DGEBA/MTHPA /BDMA/PEK-C			DGEBA/MTHPA /BDMA/PES		
PEK-C/ [η] (dL/g)	$Ea(ps)$/kJ .mol^{-1}	R	PES [η] /(dL/g)	$Ea(ps)$/kJ .mol^{-1}	R
0.53	74.8	0.999	0.36	77.8	0.998
0.43	77.1	0.999	0.43	72.0	0.999
0.32	76.1	0.999	0.53	73.5	0.999

Table 7. E_a *(ps)* values for DGEBA/MTHPA /BDMA/TP systems with different TP molecular size, BDMA 0.9%

5.3 Effect of cure rate

In our research, initiator or catalyst was employed to accelerate the TS matrix cure rate, the more the initiator/catalyst content, the fast the cure reaction will be. It was observed that phase morphologies will change in size or patterns during the variation of cure rate, whereas the phase separation time-temperature dependences keep similar. Shown in Figure 10 are the linear correlations of $\ln t_{ps}$ vs. $1/T$ for DGEBA/MTHPA/BDMA/PEK-C system containing different level of initiator BDMA and AroCyL10/PEK-C system with different catalyst content of $Cu(AcAc)_2$. Since the initiator/catalyst contents are very limited, so the thermodynamics of the mixtures is supposed to approximately the same. According to the cure reaction mechanism, the higher the initiator or catalyst levels, the faster the consumption of the TS monomers, and the earlier the systems will be driven into the unstable region. Although the phase separation times are different for composites with different level of initiator/catalyst while the parallel layouts of curves in both of the two diagrams indicate the similar E_a *(ps)* values as summarized in Table 8, even though the morphologies change with variation of initiator levels (Cui et al, 1997; Montserrat et al, 1995).

The cure reaction activation energy E_a of systems with different level of initiator/catalyst was determined by Kisinger approach as shown in Table 8 as *Ea*. It was observed that when initiator content is ≥0.9%, cure reaction activation energy *Ea* values are similar in DGEBA/MTHPA/BDMA/PEK-C systems of different BDMA content. This is different from Montserat's finding (Montserat et al, 1995), which said that *Ea* generally decreased with increase of initiator level. It is presumably because the initiator content employed in our present systems is higher enough to catalyze the chain-wise copolymerization of DGEBA and MTHPA during the temperature scan, and the higher *Ea* values in Montserat's systems probably arise from the uncatalyzed copolymerization for the deficiency of initiator BDMA. Similar results were observed in AroCyL10/Cu(AcAc)$_2$/PES systems.

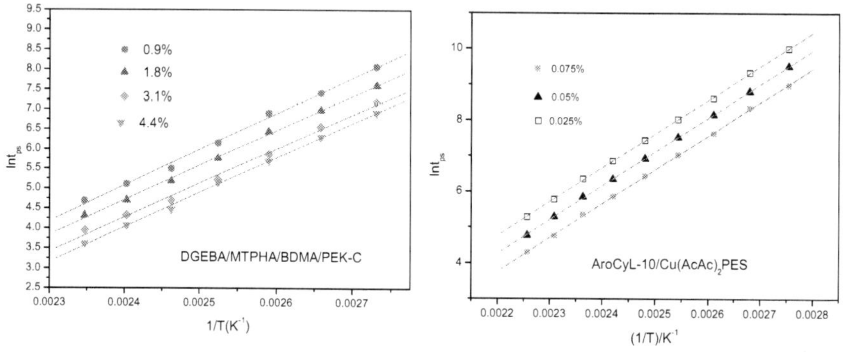

Fig. 10. lnt_{ps} *vs.* *1/T* for DGEBA/MTHPA/BDMA/PEK-C10phr and AroCyL10/Cu(AcAc)$_2$/PES10phr with different initiator content (based on OM)

DGEBA/MTHPA/BDMA/PEK-C10phr					AroCyL10/Cu(AcAc)$_2$/PES10phr		
BDMA content/ wt.%	*Ea(ps)*/ kJ .mol^{-1}	R	*Ea*/ kJ .mol^{-1}	R	Cu(AcAc)$_2$ content/wt.%	*Ea(ps)*/ kJ .mol^{-1}	R
0.9	74.8	0.999	73.8	0.999	0.025	78.8	0.999
1.8	72.5	0.999	73.9	0.999	0.05	78.5	0.999
3.1	71.4	0.999	n.d.	n.d.	0.1	77.8	0.999
4.4	72.3	0.999	74.1	0.999	n.d.	n.d.	n.d.

Table 8. *Ea (ps)* for Epoxy/hardener/TP and Cyanate ester/TP systems with different initiator /catalyst content

5.4 Cure activation energy barrier and chemical environment effects on *Ea(ps)*

As depicted in Section 5.2 and 5.3 that *Ea(ps)* values are independent on the cure rate and also irrespective to the TP content and TP molecular size in the usual systems as far as TPs are used in minor part as toughening agents. But *Ea(ps)* varies with the cure activation energy barrier *Ea* and chemical environments, both of which will be discussed in the following subsections.

5.4.1 Cure path and cure activation energy effects on *Ea(ps)*

Shown in Figure 11 and Table 9 is the phase separation time/temperature relations in the range of 120-200℃ for systems of BMI/DBA/PEK-C10phr and BMI/DBA/PES 10phr respectively, together with the gel time/temperature data. It can be seen that the slope of Arrhenius plot of the phase separation and gelation in the 120-170℃ range is lower than that in the 180-200℃ range in both of the two systems. Correspondingly, the phase separation activation energy E_a *(ps)* values in the range of 120-170 ℃ are lower than those in the range of 180-200℃. The change of temperature dependency on the phase separation times can be attributed to the complexity of the cure reaction between BMI and DBA. Although lots attempts have been made to elucidate the curing mechanism of BMI/DBA systems, these have been impeded by the complexities of the reaction multiple paths of the reaction. The following reaction types have been proposed to be involved in the curing process: ENE, Diels–Alder, homopolymerization, and alternating compolymerization. Allyl phenol compounds are expected to co-react with BMI to give linear chain extension by an ENE-type reaction in lower temperature range of 120-170℃ and show a lower cure activation energy barrier, and this is followed by a Diels–Alder reaction at a high temperature range of 180-200℃ with bigger cure activation energy (Mijovic& Andjelic,1996; Shibahara et al, 1998; Rozenberg et a,2001; Xiong et al, 2003).

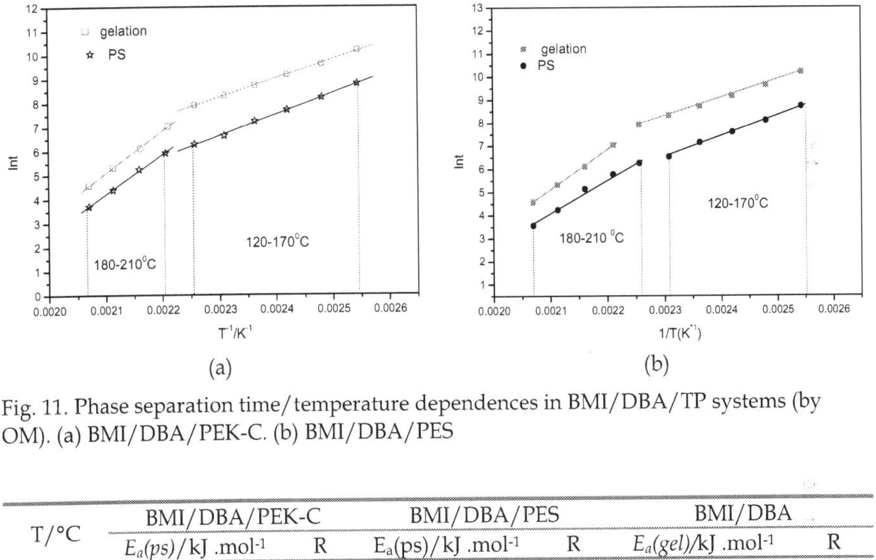

(a) (b)

Fig. 11. Phase separation time/temperature dependences in BMI/DBA/TP systems (by OM). (a) BMI/DBA/PEK-C. (b) BMI/DBA/PES

T/°C	BMI/DBA/PEK-C		BMI/DBA/PES		BMI/DBA	
	$E_a(ps)$/kJ .mol^{-1}	R	$E_a(ps)$/kJ .mol^{-1}	R	$E_a(gel)$/kJ .mol^{-1}	R
120~170	73.7	0.999	63.4	0.999	54.3	0.999
180~210	136.0	0.999	130.1	0.999	120.3	0.999

Table 9. Phase separation *Ea(ps)* values of BMI/DBA/TP 10phr systems

Systems	Initiation	*Ea(ps)*/ kJ .mol^{-1}	R	*Ea*/ kJ .mol^{-1}	R
DGEBA/MTHPA/PES10phr	Initiator 0.0%	83.9	0.999	75.5	0.999
	Initiator 0.9%	77.8	0.998	73.8	0.999
DGEBA/MTHPA/PEI10phr	Initiator 0.0%	82.3	0.999	88.3	0.999
	Initiator 0.9%	59.0	0.999	73.8	0.999

Table 10. *Ea(ps)* values for DGEBA/MTHPA/PES and DGEBA/MTHPA/PEI systems without and with BDMA

The effects of cure reaction energy *Ea(ps)* were also observed in DGEBA/MTHPA/PES and DGEBA/MHTPA/PEI systems as displayed in Table 10, wherein the initiator BDMA was employed or not. The reactions of DGEBA and MTHA with or without initiator BDMA are in different path. As reported that DGEBA and MTPHA reacts through the chain mode in the presence the catalyst such as tertiary amine BDMA with low activation energy compared to the stepwise polymerization of DGEBA and MTHPA (Fisch et al,1956; Fisch,1960). As shown in Table 10, both the DGEBA/MTHPA/PES and DGEBA/MTHPA/PEI systems without initiator BDMA show higher cure activation energy *Ea* values than the systems containing 0.9%BDMA, accordingly higher Ea(ps) values were observed. It should be pointed that in DGEBA/MTHPA/PEI system which shows UCST phase behavior, the *Ea(ps)* of composition without catalyst BDMA is much higher than the composition with BDMA. The cure activation energy changes in the range of ΔEa=14.5k J .mol^{-1}, while phase separation activation energy show the discrepancy of $\Delta Ea(ps)$=22.3k J.mol^{-1}, which is much higher than the former. It can be assumed that the significant increase of *Ea(ps)* of DGEBA/MTHPA/PEI composition is not only generated from the change of chemical reaction energy barrier *Ea*, while part of the increase of $\Delta Ea(ps)$ may come from the increase of the cure reaction temperature. In DGEBA/MTHPA/PEI composition, because the absence of initiator BDMA, the cure temperature has been elevated quite lot for cure efficiency, as was reported, DGEBA/PEI is a system of UCST phase diagram (Bonnaud et al, 2002), the elevation of cure temperature will increase the quench depth which possibly causes the increase of *Ea(ps)*. The physical scheme of the correlation and even similar value range implies the CIPS induced by the curing.

5.4.2 *Ea(ps)* dependence on stoichiometry in UCST TP/TS/hardener systems

Besides the chemical reaction energy barrier dependence of *Ea(ps)*, it was observed that the *Ea(ps)* values also vary with the chemical environments, like stoichiometric balance, TP and TS monomer structures. We will discuss how the chemical environment factors will alter the phase separation activation energy in details. In this subsection, we only talk about the phase separation time-temperature dependence in TP modified TS systems with UCST type phase diagram, which can be explained quite well with the interaction energy density parameter in our research scope.

Shown in Figure 12 and Table 11 are the time-temperature dependence variation in DGEBA/DDM(DDS)/PEI10phr and DGEBA/MTHPA/BDMA/PEI10phr systems with different chemical stoichiometric ratio *r*. As can be seen in DGEBA/DDM/PEI system which shows UCST type phase behavior (Bonnaud et al, 2002), the phase separation activation energy values *Ea(ps)* of it increase with *r*, when *r*=1.5 the system even shows no phase separation. While in DGEBA/DDS/PEI system, *Ea(ps)* changes with *r* in an opposite way. In view of chemical reaction energy barrier, E_a does not change with the stoichiometric ratio *r* for both of the DGEBA/DDM/PEI and DGEBA/DDS/PEI systems as displayed in Table 11 by value of *Ea*. The changes of *Ea(ps)* possibly originate from the variation of chemical environments with the change of stoichiometric ratio, e.g. the miscibility between components is altered with the increase of *r* values, which in turn influence the *Ea(ps)* substantially. We calculated the interaction energy density *B* values for different combination as shown in Table 12. It can be seen that the miscibility of PEI with DDS is worse than that it with the cured DGEBA/DDS. So it can be hypothesized that the more of

DDS, the poorer the miscibility between TP and cured TS and the easier for the onset of phase separation with lower *Ea(ps)*. The inverse scenarios were observed in DGEBA/DDM/PEI systems, wherein DDM is unfavorable for demixing of the components of TP and TS, DDM shows better affinity to PEI than the cured DGEBA/DDM matrix, so the more the DDM content, the better the compatibility between the components and the bigger *Ea(ps)* value it exhibits. Similar phenomena were observed in DGEBA/MTHPA/PEI systems which can also be explained in the frame of interaction energy density parameter.

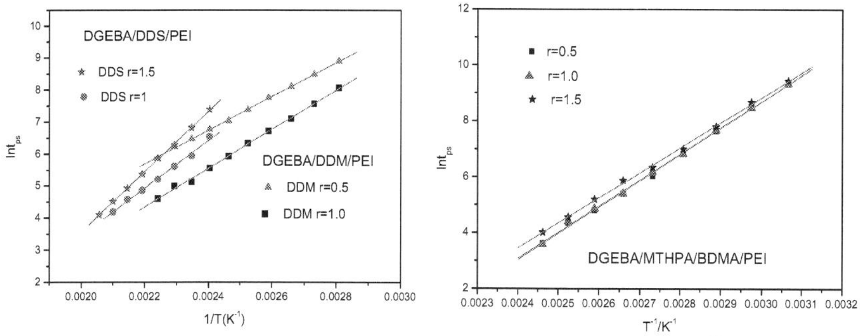

Fig. 12. Plots of lnt_{ps} vs. *1/T* for TP/ TS/hardener systems with different stoichiometry *r* values

r	DGEBA/DDM/PEI *Ea(ps)*/ kJ .mol-1	DGEBA/DDM/PEI *Ea(gel)*/ kJ .mol-1	DGEBA/DDS/PEI *Ea(ps)*/ kJ .mol-1	DGEBA/DDS/PEI *Ea(gel)*/ kJ .mol-1	DGEBA/MTHPA/BDMA/PEI *Ea(ps)*/ kJ .mol-1	DGEBA/MTHPA/BDMA/PEI *Ea(gel)*/ kJ .mol-1
0.5	43.8	46.3	no demixing	61.1	56.3	51.2
1.0	50.6	44.6	78.6	58.1	56.2	51.6
1.5	No ps	44.7	62.2	59.7	59.70	51.7

Table 11. Values of *Ea (ps)* for different TP/TS/hardener systems with different stoichiometric ratios *r*

B/J.cm-3	DGEBA/DDM	DGEBA/DDS	DGEBA/MTHPA	DDM	MTHPA	DDS	AroCyL-10
PEI	5.61	0.12	6.80	3.60	0.50	2.48	3.17
PEI1	3.13	0.89	4.03	1.68	0.012	4.72	5.67

Table 12. Interaction energy density *B* values between different combinations

5.4.3 TP and TS structure effects on *Ea(ps)* in UCST type TP/TS/hardener systems

Besides the chemical stoichiometric balance, TP or TS structure variation all take substantial effects on the phase separation time-temperature dependence. As shown in Table 13, the phase separation activation energy *Ea(ps)* varies seriously with the poly(ether imide) structures changing, either in epoxy/amine and epoxy/anhydride matrices, or in cyanate ester systems. PEI and PEI1 have different chemical structures as shown in Table 1, both of them show different miscibility with matrices of DGEBA/DDM and DGEBA/DDS respectively. In DGEBA/DDM matrix, PEI1 show higher affinity to the matrix than PEI does, so it can be understood that even no phase separation was observed in DGEBA/DDM/ PEI_1 for the excellent compatibility; while in DGEBA/DDS matrix, miscibility trend reverses, PEI is more favorable to phase mixing than PEI_1 as denoted by the higher *Ea(ps)* value of 78.6 k J $.mol^{-1}$. Similar phenomena were observed in the matrix of DGEBA/anhydride/PEI and AroCyL10/PEI respectively. In AroCyL10/Cu(AcAc)2/PEI1 system PEI1 even can't be dissolved in the matrix homogenously within the experimental condition for the poor miscibility of the TP and AroCyL10 monomers, the bigger interaction energy density parameter *B* in Table 15 gives the reasonable interpretation. In all these mixtures, the variations of *Ea(ps)* because of PEI structure dissimilarity can be well interpreted by the interaction energy density theory.

TP	Ea(ps) /kJ .mol-1	R	Ea(gel)/ kJ .mol-1	R	Ea(ps) /kJ .mol-1	R	Ea(gel) /kJ .mol-1	R
	DGEBA/DDM matrix				DGEBA/DDS matrix			
PEI1	no ps	-	51.6	0.999	61.2	0.999	62.8	0.999
PEI	50.6	0.999	51.0	0.999	78.6	0.999	62.5	0.999
TP	DGEBA/MTHPA matrix				AroCyL10/Cu(AcAc)2 matrix			
PEI1	72.6	0.999	52.7	0.999	Immiscible	/	60.9	0.999
PEI	59.7	0.999	51.6	0.999	47.9	0.999	61.2	0.999

Table 13. Phase separation time-temperature dependence of TS/hardener/PEI systems with different TP structure

Epoxy monomers	$Ea(ps)/kJ.mol^{-1}$	R	$Ea/kJ.mol^{-1}$	R
E56	54.7	0.999	51.4	0.999
E54	52.6	0.999	/	/
E51	52.4	0.999	51.6	0.999
E44	44.3	0.999	/	/
E39	40.3	0.999	51.2	0.99
E51/ E44(1:1,ratio by weight)	48.7	0.999	/	/
E51/E39 (1:1,ratio by weight)	40.1	0.999	/	/
E51/ E31(4:1,ratio by weight)	36.2	0.999	52.1	0.999

Table 14. *Ea (ps)* for epoxy/DDM/ PEI systems with different epoxy monomer structures

The phase separation time-temperature dependencies also changed with the variation of TS monomer structures, as shown in Table 14, with the increase of epoxy molecular weight, *Ea(ps)* changes substantially. Here all the epoxy monomers have the structures referring to Table 1, only differing in the number of repeating unit n, the numbers beside "E" in Table 14 donating the epoxide value of the epoxy monomer, like E54 means monomer contains 0.54mol epoxide functional group per 100g DGEBA., The smaller epoxide value, the bigger the monomer size. As can be seen that the phase separation activation energy Ea(ps) decreases with the increase of molecular size of epoxy monomer DGEBA. Interaction energy density parameter *B* values between the cured DGEBA/DDM matrix and PEI were calculated and summarized in Table 15. As was shown that the growth of DGEBA monomer size give higher *B* value implying the widening of miscibility mismatch between TP and TS components. This is apparently contradictory to the finding in above section wherein TP content and size show limited effect on *Ea(ps)* values. It is true that the increase of DGEBA molecular size will change the phase diagram location as were observed, but except the DGEBA size effect on phase behavior, there is another favoring factor for phase separation is the volume fraction change of cure agent DDM because of the increase of DGEBA size. Bigger DGEBA gives higher molar volume fraction which means small volume fraction remained for cure agent DDM. DDM is a component which is very favorable of phase mixing, the less of DDM volume is, the worse the miscibility of TP and TS will be.

B($J.cm^{-3}$)	E56/DDM	E54/DDM	E51/DDM	E44/DDM	E39/DDM	E31/DDM
PEI	5.53	5.56	5.61	5.74	5.83	10.71

Table 15. Change of interaction energy density parameter *B* for the variation of epoxy monomer structures

5.4.4 Stoichiometry and TP and TS structure effects on *Ea(ps)* in LCST type TP/TS/hardener systems

The interaction energy density theory works well in rationalization of the phase separation time-temperature dependence on the chemical environment change: ether coming from stoichiometric imbalance, or TP and TS structures in the TP/TS systems with UCST phase behavior. But such theory is not universal in TP/TS systems with LCST phase diagrams.

Shown in Table 16 are the phase separation activation energy *Ea(ps)* values of DGEBA/DDM/PES (Bonnaud et al, 2002) and DGEBA/MTHPA/PES with different stoichiometry *r* both of which show LCST type phase diagram. As was shown that both of *Ea(ps)* values increase with the descending of *r* value. The interaction energy density parameter *B* of PES/harder and PES/(cured TS) are calculated and displayed in Table 18. It was found that both of the two hardeners of DDM and MTHPA show higher affinity to PES than the corresponding cured resin, while the experimental observations did not support the theoretical prediction, the higher the hardener content was, the smaller the phase separation energy barrier appeared. The apparent paradoxical phenomenon comes from the limitation of the present thermodynamics in depicting polymer mixture phase separation process, which will be explained in following content within this section.

	DGEBA/DDM/PES				DGEBA/MTHPA/PES			
r	Ea(ps) /$kJ.mol^{-1}$	R	Ea(gel) /$kJ.mol^{-1}$	R	Ea(ps) /$kJ.mol^{-1}$	R	Ea(gel) /$kJ.mol^{-1}$	R
1.5	53.7	0.999	44.7	0.999	72.8	0.999	70.2	0.999
1.0	58.6	0.998	44.6	0.998	77.8	0.999	70.3	0.999
0.5	62.4	0.998	46.3	0.998	78.3	0.999	70.7	0.999

Table 16. *Ea(ps)* variations in DGEBA/DDM/PES and DGEBA/MTHPA/PES systems with different *r*

Besides the stoichiometric effects on the variation of phase separation activation energy *Ea(ps)* in various TP modified TS systems which show LCST phase diagrams, the chemical environments effects arising from TP and TS structure alterations were also studied. Here we mainly talk about some representative systems, like DGEBA/DDM/TP, DGEBA/MTHPA/BDAM /TP and Cyanate ester system of AroCyL10/TP (Hwang et al, 1999) as in Table 17, wherein TP structure varies in repeating units.

Based on interaction energy density parameter calculation between different combinations as shown in Table 18, it can be seen that in DGEBA/DDM/TP systems, PEK-S shows highest affinity to the matrix, then PSF and PES show dropping affinity to matrix in sequential. While the measured *Ea(ps)* values have not shown the corresponding theoretical sequence. It can be seen that DGEBA/DDM/PES-C with highest miscibility did show biggest phase separation energy barrier, e.g. no phase separation was observed upon completion of the cure. But DGEBA/DDM/PES shows a higher *Ea(ps)* value than that of DGEBA/DDM/PSF system which is contrary to theoretical prediction. Inconsistence was also observed in other systems, e.g. DGEBA/MTHPA/PEK-C systems, *Ea(ps)* doesn't show a monotonic relationship with the calculated *B* value.

All these experimental and theoretical discrepancies arise from the limitation of the present cohesive energy theory of interaction energy interaction density. Indeed, volume fraction change during cure, strong hydrogen bonding and strong polar interaction are not included in our calculation. These effects could be critical for the phase behavior, especially in the LCST systems, so further theoretical and experimental efforts are to be made. Up to now, all the theoretical analysis works only well in UCST systems where hydrogen bonding and other special energy effect are absence. But for the LCST systems, as have been found in this subsection, most of the experimental results contradict the theoretical hypothesis.

TP type	DGEBA /DDM/TP		DGEBA /MTHPA/TP		AroCyL-10/TP	
	Ea(ps) /kJ.mol^{-1}	Ea /kJ.mol^{-1}	Ea(ps)/kJ.mol^{-1}	Ea /kJ.mol^{-1}	Ea /kJ.mol^{-1}	Ea(ps) /kJ.mol^{-1}
PES	60.1	50.7	60.1		78.5	60.8
PSF	55.9	51.6	55.9		70.3	65.2
PES-C	No PS	51.9	No PS		/	/

Table 17. *Ea(ps)* variation for the change of TP structure in matrix of TS/hardener/TP mixtures

B(J.cm^{-3})	DGEBA/DDM	DDM	DGEBA/MTHPA	MTHPA	AroCyL-10
PES	16.56	12.68	19.00	9.38	31.2
PSF	6.60	/	7.02	/	47.6
PES-C	3.13	/	3.42	/	/

Table 18. Calculated *B* for PES, PSF and PES-C with TS components

5.4 Time-temperature transformation diagram with phase separation

As was known that all the processing of thermosetting composites are carried out in wide time/temperature space as presented by TTT (time-temperature-transformation) diagram (Enns & Gillham,1983;Gillham,1986). Phase separation usually occurs before the chemical gelation, which is always accompanied by the change of viscosity or modulus as shown by Figure 2 in Section 3. Following phase separation is the chemical gelation of TS rich phase, which retards the resin flow and limits the coarsening of the dispersed phase domain. Upon vitrification, curing ceased totally because the freezing of the component chain.

For TP modified TS processing, the cure time-temperature routine has a great impact on the morphology generated during cure further to the final material properties (Inoue, 1995; Williarms, 1997). To get TP/TS material with desired structure and properties, it is necessary to design the cure time-temperature processing routine. The time-temperature dependence of gel and vitrification is well illustrated by the kinetic approach for the systems with unambiguous cure kinetics (Wisanrakkit & Gillham,1990). It is desired to have the phase separation zone defined in the TTT diagram for TP/TS composite processing. So far only schematic diagram appeared in literatures (Williams, et al. 1997). This situation is obviously related to the issue of determination of phase separation time in broad temperature range.

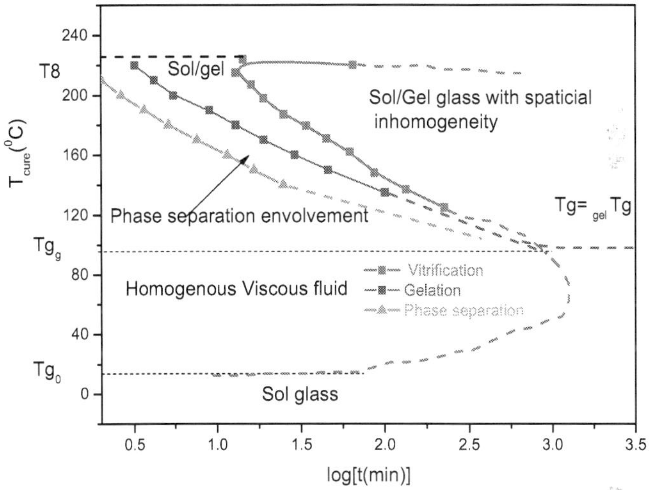

Fig. 13. TTT diagram depicting occurring times for phase separation, chemical gelation and vitrification for Epoxy/DDS/ PEK-C 15phr system

Based on our efforts on the phase-separation time-temperature dependence relationship and the TS matrix cure kinetics, it is now convenient to build the Time-Temperature Transformation Diagram (TTT) accompanied with phase separation process (TTT-PS diagram). Figure 13 displays the example of experimental determined TTT-PS diagram, which defines the main events taking place during the curing of TP modified TS with the presence of CIPS in whole time-temperature window. Obviously the TTT-PS diagram may be of great importance for composites processing where morphology and structure control is of interests. It is shown in Figure 13 that in the measured Epoxy/DDS/PKE-C system with a LCST phase behavior, the phase separations take place before chemical gelation throughout the cure temperature range employed, and the phase separated structure was even observed at temperature where the epoxy matrix started degrading with volatiles coming out. In some TP/TS mixture, e.g. UCST systems, phase separations could only take place in certain time-temperature space within the up-merging curves of phase separation and gelation. Even in this area PS is not guarantied, because of the PS dynamics at early critical stages and the pattern developing dynamics. In Figure 13, the phase separation and gelation time-temperature curves in the semi-log coordinator at low temperature range are extrapolated in a convergent direction intentionally to depict to what will happen at low temperature. It can be assumed theoretically that the phase separation time-temperature curve and the gelation time-temperature will converge below certain temperature. This is because our present mixture shows LUST phase behavior. The decrease of temperature is favorable of phase mixing which possibly will contribute to a higher phase separation activation energy; on the other hand, the cure reaction will become difficult because the restrain on component mobility generated by the high viscosity accompanying the temperature decrease.

6. Morphology evolvement in PS and dynamic theory prediction

6.1 Dynamic theory for morphology evolvement

The thermodynamic theory reveals the possibility of phase separation (PS), the phase separation possible mechanism like spinodal & bimodal decompositions, and the conditions of PS emerging, while it does not answer how the PS pattern develops, e.g. the phase separation dynamics. To understand the phase pattern evolvement (for example in Figure 7) and control the morphology in processed product, the phase separation dynamics has been intensively studied in the past three decades from both experimental and theoretical viewpoints(Onuki,1986; Onuki & Taniguchi 1991; Tanaka, 1997; Onuki & Taniguchi 1997). Experimental observations of CIPS disclosed spectacular variable schemes of morphology changes depending on the composition and process parameters. Good news is that owing to the advances of condensed matter physics a unified scheme became possible to explain the complex phenomena. From the concept of dynamic universality of critical phenomena, phase separation phenomena have been classified into various theoretical models. For example, phase separation in solids is known as the "Solid Model or Model B" (Hohenberg & Halperin, 1977), where the local concentration can be changed only by material diffusion, while phase separation in fluids is known as the "fluid model or model H" by both diffusion and flow (Doi&Onuki, 1992). It has been established that within each group the behavior is universal and does not depend on the details of the material. The classic theories consider the same dynamics for the two components of a binary mixture, which we call "dynamic symmetry". However, such an assumption of dynamic symmetry is hardly valid in "complex fluids". For the example of TP/TS mixtures, their molecular mass, diffusivity and mobility are very different. For these "dynamic asymmetric" systems one needs to consider the interplay between critical dynamics and the slow dynamics of polymer itself which was in so called "Viscoelastic Model" (Tanaka, 1997).

Some researches have been focused on the CIPS process starting from homogenous solution using Onuki's Elastic solid model (Zhang, 1999) or "two fluid" model or Tanaka's viscoelastic model (Tanaka, 1997). Which model should one choose depends what dominant physical parameters are involved in the system. For example, at the early stage of spinodal phase separation of high viscous mixture, diffusion is dominant, so elastic solid model may work well enough, while in late stage under interfacial tension at gradually clearer phase border the drop merging becomes important and solid model may underestimate the pattern coarsening. Simulation is often a necessary approach to "image" the events take place in the black box where we can't tough directly by any means for its complexity, while numerical solutions work successfully.

In our present work, the modified Elastic Model of Onuki (Onuki& Taniguchi 1991) was employed, the phase separation dynamical equation was numerically solved, using the so-called cell dynamical system presented by Oono and Puri (Oono & Puri 1988).

The free energy functional of the system F is given by (Onuki, 1997; Zhang, 1999):

$$F = \int d\mathbf{r}[f(\phi) + \frac{1}{2}|\nabla\phi|^2 + a\phi\nabla \bullet \mathbf{u} + \frac{1}{2}K|\nabla \bullet \mathbf{u}|^2 + \mu Q] \tag{12}$$

Wherein $f(\phi)$ $+1/2|\nabla\phi|^2$ is the free energy of unit volume in the Flory-Huggins average lattice theory or Ginzberg-Landau theory, a is a parameter represents the coupling constant of concentration gradient and elastic field, u is the deformation vector. In linear viscoelastic region, the bulk elastic energy originates from the bulk modulus K and symmetric conformation tensor Q, which are irrespective to volume fraction Φ, and shear energy comes from the shear modulus μ and symmetric conformation tensor Q:

$$Q = \frac{1}{4}\sum_{i,j}[\nabla_i u_j + \nabla_j u_i - \delta_{ij}\frac{2}{d}\nabla \bullet \mathbf{u}]^2 \ , \ \nabla_i = \frac{\partial}{\partial x_i} \tag{13}$$

Wherein d is the spatial dimension μ and concentration ϕ is supposed to be linear as:

$$\mu = \mu_0 + \mu_1\phi \tag{14}$$

Based on the force equilibrium requirement of $\frac{\delta F}{\delta u_i} = 0$ we can use the following equation to replace u in equation 12, at the same time we can get the relationship of concentration and elastic energy:

$$\delta u_i = -(\alpha / K_L)\frac{\partial \omega}{\partial x_i} \tag{15}$$

$$\nabla^2\omega = \phi - \overline{\phi} \tag{16}$$

Where $\overline{\phi}$ is the spatial average concentration, K_L can be correlated to bulk modulus through equation 17：

$$K_L = K + 2(1 - 1/d)u_0 \tag{17}$$

Under infinitesimal deformation near equilibrium, the symmetric conformation tensor Q can be extended based on δu ， and only first order approximation is left.

$$\hat{Q} = (\alpha / K_L)^2 \sum_{i,j} [\nabla_i \nabla_j \omega - \frac{1}{d}\delta_{ij}\nabla^2 \omega]^2 \tag{18}$$

For simplicity, all the constant part in the free energy, the free energy functional can be expressed as

$$F = \int d\mathbf{r}[f(\phi) + \frac{1}{2}|\nabla\phi|^2 + \mu_1 \phi \hat{Q}] \tag{19}$$

And with equation 18, we can construct the TDGL (time dependent Ginzberg Landau) equation :

$$\begin{aligned} \frac{\partial \phi}{\partial t} &= M\nabla^2 \frac{\delta F}{\delta \phi} = M\nabla^2 [\frac{\partial f}{\partial \phi} - \kappa \nabla^2 \omega \\ &+ G_E \hat{Q}] + 2G_E \sum_{i,j} \nabla_i \nabla_j \phi [\sum_{i,j} \nabla_i \nabla_j \phi - \frac{1}{2}\delta_{ij}(\phi - \bar{\phi})] \end{aligned} \tag{20}$$

Where in

$$g_E = \mu_1 (\alpha / K_L)^2 \tag{20a}$$

is the modulus of TS phase, it changes with the cure time for the increase of TS size, for most of the TS monomers reaction through the second order reaction. According to the chemorheology of curing, one can get the time-temperature dependence of dynamic viscosity change. Therefore, in our simulation, we use the expression $g_E \sim$ (TS conversion)

$$1 / ((1 - \exp(-\mathrm{Ea} / \mathrm{RT}) * \mathrm{t}) \tag{20b}$$

The Ginzberg-Landau free energy is simplified as following:

$$\frac{\partial f}{\partial \phi} - \kappa \nabla^2 \phi = -\phi + \phi^3 - \nabla^2 \phi \tag{21}$$

By numerical calculation of governing equation of 16, 20 and 21 we can get the qualitative information about what happened in the intermediate and later stager of CIPS using the modified "Solid Elastic Model" wherein modulus mismatch between TP and TS varied along time. Shown in Figure 14 is a simulation result for the systems with "Viscoelastic symmetric" modulus. The volume fractions of TP component is φ=0.5. As can be seen in the "dynamic viscelastic symmetric" system, bi-continuous phase was observed at early stage and some coarsening of the networks appear with time until at last TP network morphologies freeze , because of full gelation of TS phase

In some TP/TS/Hardener systems when the TP molecular is not high enough or TS monomers show high hydrodynamic effect upon phase separation, second phase separations were always observed. By the present "Elastic Solid Model", such scenario can't be incorporated, wherein in the "Two Fluid Model" coupled with chemical reaction works better. But for the composites material fabrication, the final morphology though too much interface sharpening and phase coarsening via flow may not a favorite process.

In the real processing of composites material during fabrication of large parts, like composite aircraft wind and fuselage, incomplete diffusion with elastic misfit between the components does exist, wherein the TP and TS are interlaminated during RTM (resin transfer molding). Based on ex situ toughening concept, the thin TP layers are periodically interleaved the TS/graphite prepreg plies, the respective TP and TS components were separated. After the wet prepregs are laminated, the interaction between TP and TS phase and the diffusion couples with the cure reaction and phase separation start (Yi,2006;Yi et al, 2008; Yi & An,2008; Yi, 2009). In such delicate, hard to monitor process, the computer simulation can play role. We will discuss such simulation process with modified Onuki "Elastic Solid Model" in subsequent section.

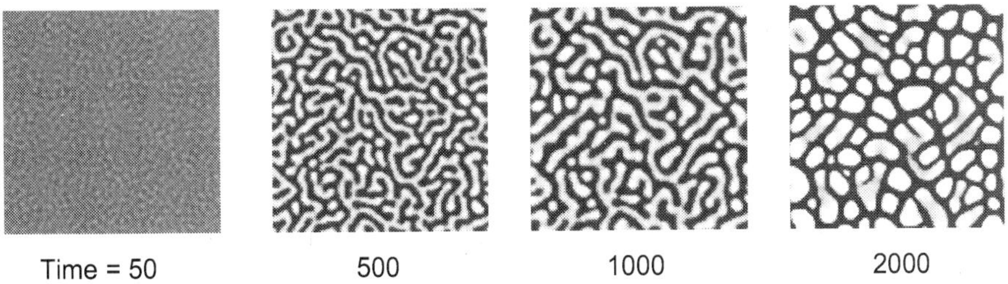

Fig. 14. Morphology evolvement of two components mixture with symmetric moduli. t donating the dimensionless time in the temporal space. Volume fraction is 50%

6.2 Spatially inhomogeneous structure with concentration gradient

In Section 6.1 we simulate what happens in the *in situ* phase separation process, wherein the phase separated structures are distributed spatially evenly in the matrix, phase inversion mechanism was clearly displayed. Here, we will use the same phase separation thermodynamic and similar kinetic frame to monitor the events happened in *ex situ* toughening process, where interleaved TP and TS layered are layout periodically in spatial direction. By a similar two dimension CDS process , a linear concentration gradient is assumed along y-coordinate, our computational space is graded in square of size $0<x<L$ and $0<y<L$, wherein L is the dimensionless square grid length, the concentration gradient is shown as $\phi(y)=\frac{1}{L}*y$ and $\frac{d\phi}{dx}=0$, where in ϕ is the volume fraction of minor component TP.

Other thermodynamics and phase separation dynamics descriptions keep the same as Section 6.1 wherein in situ phase separation process is calculated by the CDS approach. In the present system, Shown in Figure 15 are simulated morphology evolving process with varying TP and TS modulus mismatch value of *Ge*. It can be observed that the higher the asymmetry is, the more inverted structures formed in the interface between TP and TS matrices. In a *Ge*=100 system, bi-continuous and inversion structures appear throughout the direction perpendicular to the layer plane, and also the morphology distribution varies along the same direction.

Shown in Figure 16 is the experimental SEM (Scanning Electronic Microscope) observation on the fractured interface between the epoxy BEP/DDS matrix and thermoplastic PEK-C layer. The morphology size and pattern vary periodically in direction perpendicular to ply direction. The simulation scenarios fit the experimental results qualitatively. In the future, improvements of present model with incorporation of spatial temperature ingredient and cure cycle possibly will give more clues to how the morphologies evolve during the real composite manufacture process.

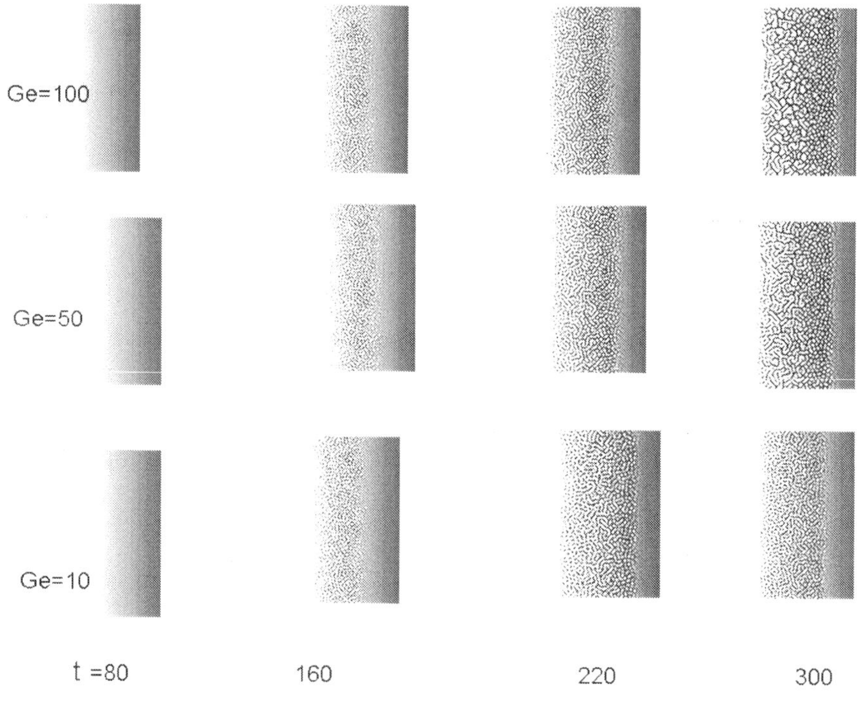

Fig. 15. Effect of modulus mismatch value Ge on the morphology distribution in direction perpendicular to ply

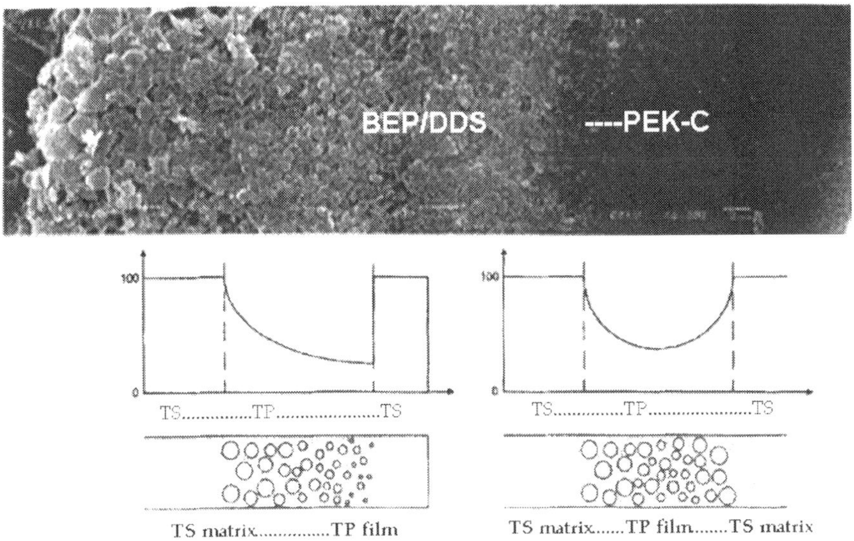

Fig. 16. Morphology distribution of an etched TP/TS/graphite composite toughened by ex-situ interlaminar process, observed by scanning electron microscope

7. Concluding remarks

In the past three decades, numerous experimental and theoretical achievements dealing with reaction-induced phase separation (RIPS) of polymer systems were driven by versatile applications in spite of the complex nature of the subject from the theoretical point of view. Cure induced phase separation (CIPS) is an important part of RIPS due to its innovative applications to composite processing, or more generally, to the innovations of multi-phase polymers. Instead of fight with interfacial tension in polyblends, CIPS processes involve the creation of phases with granular or network morphologies, the interfacial sharpening & coarsening according to the balance of diffusion and convection of viscous mobile systems, the termination of PS or the secondary PS affected by TS gelation and vitrification. It seems to be a disadvantage to allow the morphology influenced by so many factors, but it's also an advantage to have many means to create the structure we want, if we know the control rules. The intention of this review was using our practice to provide a qualitative (or partially quantitative) basis for readers to rationalize the numerous factors in CIPS process to achieve the desired morphologies and composite properties.

In the first section it was shown that: a) the purpose of modifying a high performance TS polymer composite is calling on innovative multi-scale matrix design; b) to achieve high toughness and, at the same time, high rigidity, the controllable micro-inhomogeneity including bi-continuous network with adjustable interface could be a good option for composite matrices; c) cure induced phase separation of the mixture of thermosetting resin (TS) with engineering plastics (TP) provide promising approach to do so. d) a gradient design of PS fitted in layered composite can even more optimize its performance.

To detect and study CIPS process comprehensively, one need multi measuring means, including necessary experimental innovations, which are summarized as: a) to detect the phase separation time (PS) in early stage, a patented transmission optical morphology analyzer (TOM) allows determining PS time and tracing the image evolvement of micro-phase separation at high T and long time; b) comparative studies using OM, SALS & dynamic rheology found two critical transitions: TP gelation by PS and a TS gelation, revealed the loose fractal network structure of TP at PS and TS fractal network at "critical gel states" and determined the network fractal dimensions.

Based on these "on-line" experimental approaches we were able to study the cure reaction of TP/TS with PS in broad time-temperature range. The temperature dependence of PS time was found experimentally to obey Arrhenius type rule for various TP/TS/hardener systems and PS activation energy may be defined and its relations to many compositional parameters were clarified experimentally. We were able to build the Time-Temperature-Transformation diagram including phase separation (TTT-PS) experimentally, which is of essential importance for TP/TS composite processing. There is still long way to go to answer the following questions thoroughly: why the CIPS happens? What governs the phase separated morphology development? How to predict the appearing of CIPS from the chemical formulation and processing parameters? However, some progresses in this direction can build a promising picture: a) like all mixing thermodynamics, Flory-Huggins mean field theory can be performed for CIPS. Instead using interaction parameter χ, the interaction energy density B was employed with more reasonable physical meaning and good relation with cohesive energy and, thus, chemical structures of the components; b) for the dynamically very asymmetric system, the viscoelastic phase separation theory by incorporating the reaction kinetics into time dependent Ginsburg-Landau (TDGL) equation and cellular-automata simulation can simulate the morphology development. If interlaminate diffusion is cooperated, the gradient morphology in layered composite can be simulated; c) our practice and the theoretical background are well situated in a multi-physics frame. Either thermodynamic or dynamic theory is all in mesoscopic level, with unified rules and leaving lower level chemical reactions and structures in parameters. This paradigm should work and be proved further in our efforts to build better composites.

Most of our researches were carried out under the support of National Key Basic Research Program (973 Program, 2003CB615600)

8. Reference

An, X.F. Study on Laminated-Toughened CFRPs Based Thermoset/Thermoplastic Diphase System. PhD Dissertation, (2004), Zhejiang University

Bicerano, J. (2002-08-01).*Prediction of Polymer Properties*, (3rd Edition), CRC Press, ISBN: 0824708210

Bonnaud, L.;Pascault, P. J. ; Sautereau, H. Kinetic of a Thermoplastic-Modified Epoxy-Aromatic Diamine Formulation: Modeling And Influence of a Trifunctional Epoxy Prepolymer. (2000). *Eur. Polym. J.*,Vol. 36, No.7, pp.1313-1321

Bonnet, A.; Pascault, J.P.; Sautereau, H.; Camberlin, Y. Epoxy-diamine Thermoset/Thermoplastic Blends. 2. Rheological Behavior Before and after Phase Separation. (1999). *Macromolecules*, Vol.32, No.25, pp.8524-8530

Bonnaud, L.; Bonnet, A.; Pascault, P. J.; Sautereau, H.; Riccardi, C. C. Different Parameters Controlling the Initial Solubility of Two Thermoplastics In Epoxy Reactive Solvents. (2002). *J. Appl. Polym. Sci.*, Vol. 83, No. 6, pp.1385–1396

Borrajo, J.; Reccardi, C.C.; Williams, J.J.R. Rubber-Modified Cyanate: Thermodynamic Analysis Of Phase Separation. (1995). *Polymer*, Vol.36, No.18, 3541-3547

Bucknall, C B, Gomez, C M, Quintard, I. Phase-Separation from Solutions of Poly(Ether Sulfone) in Epoxy-Resins. (1994). *Polymer*, Vol.35, No.2, pp353-359

Bucknall, C.B.; Gilbert, A. H. Toughening Tetrafunctional Epoxy Resins Using Polyetherimide. (1989). *Polymer*, Vol. 30, No.2, pp.213-217

Bucknall, C. B.; Partridge, I. K. Phase Separation In Epoxy Resins Containing Polyethersulphone. (1983). *Polymer*, 24, pp.639-644

Cadenato, A.; Salla, M. J.; Ramis, X.; Morancho, M. J.; Marroyo, M. L.; Martin, L. J. Determination of Gel and Vitrification times of Thermoset Curing Process by Means of TMA, DMTA and DSC Techniques TTT Diagram.(1997). *J. Therm. Anal.*, Vol. 49, 269-279

Chambon, F.; Petrovic, S.Z.; Macknight, J.W.; Winter, H.H. Rheology of Model Polyurethanes at the Gel Point.(1986). *Macromolecules*, Vol.19, No.8, pp.2146-2149

Chen,C.M; Hourston, J.D.; Sun,B.W. the Morphology and Fracture Behaviour of a Miscible Epoxy Resin-Polyetherimide Blend. (1995). *Eur.Polym.J.* Vol.31,No.2,pp199-201

Cheng, Q.F.;Fang, Z.P.;Xu, Y.H.;Yi, X.S. Morphological and Spatial Effects on Toughness and Impact Damage Resistance of PAKE-toughened BMI and graphites. (2009). *Chinese Journal of Aeronautic*, Vol.22，pp87-96

Cho, B.J J. ;. Hwang, W.; Cho, K. ; An, H. J. Park, E.C. Effects of morphology on toughening of tetrafunctional epoxy resins with poly(ether imide). (1993). *Polymer,* Volum.34, No.23, pp.4832-4836

Cui, J.; Chen, W.J.;Zhang, Z.C.; Li, S.J. Studies on the Phase Separation of Polyetherimide-Modified Epoxy Resin 1.Effect of Curing Rate on the Phase Structure. (1997). *Macromol. Chem. Phys.*, Vol.198, No6, pp.1865-1872

Doi,M; Onuki,A. Dynamic Coupling Between Stress And Composition In Polymer Solutions And Blends.(1992). *J. Phys. II France* 2, pp.1631-1656

Eloundou, P.J.; Gerard, F.J.; Harran, D.; Pascault, P.J. Temperature Dependence of the Behavior of a Reactive Epoxy- Amine System by Means of Dynamic Rheology. 2.High-Tg Epoxy-Amine System.(1996). *Macromolecules,* Vol.29, No.21, pp.6917-6927

Enns,J. B.; Gillham, J. K. The Time-Temperature-Transformation (TTT) Cure Diagram: Modeling the Cure Behavior of Thermosets.(1983). *J. Appl. Polym. Sci.*,Vol. 28, No.8, pp.2567-2591

Fischer, W.; Hofmann, W.; Kaskikallio, J. the Curing Mechanism of Epoxy Resins. (1956). *J. Appl. Chem.*, Vol.6, No.10, pp.429-441

Fischer, W. Polyesters from Expoxides and Anhydrides. (1960). *J. Polym. Sci.*,Vol.44, No.173, pp.155-172

Fornasiero, F.; Olaya, M. M.; Wagner, I.; Brüderle, F.; Prausnitz, M.J. Solubilities of Nonvolatile Solutes in Polymers from Molecular Thermodynamics. (2002). *AIChE J.*, Vol.48, No.6, pp.1284-1291

Gan, W.J.; Yu, Y.F.; Wang, M.H.; Tao, Q.S.; Li, S.J. Viscoelastic Dffects on the Phase Separation in Hermoplastics-Modified Epoxy Resin. (2003). *Macromolecules*, Vol.36, No20, pp.7746-7751

Gillham, K.J. Formation and Properties of Thermosetting and High Tg Polymeric Materials. Polym. Eng. Sci., Vol.26, No.20, 1429-1433

Girard-Reydet, E.; Sautereau, H.; Pascault, J.P.; Keates, P.; Navard, P.; Thollet, G.; Vigier, G. Reaction-Induced Phase Separation Mechanisms in Modified Thermosets. (1998). *Polymer*, Vol. 39, No.11, pp.2269-2279

Girard-Reydet, E.; Vocard, V.; Pascault, P. J.; Sauterau, H. Polyetherimide-Modified Epoxy Networks: Influence of Cure Conditions on Morphology and Mechanical Properties. (1997). *J. Appl. Polym. Sci.*, Vol. 65, No.12, pp.2433–2445

Grillet, A.C.; Galy, J.; Pascault, J.P. Influence of a 2-Step Process and of Different Cure Schedules on the Generated Morphology of a Rubber-Modified Epoxy System Based on Aromatic Diamines.(1992).*Polymer*, Vol.33, No.1, pp34-43.

Halley, J.P & Mackay, E.M. Chemorheology of Thermosets-An Overview. (1996). *Polym. Eng. Sci.*, Vol. 36, No. 5 pp.693-609

Hedrick, L .J.; Yilgör, I.; Wilkes, L. G.; McGrath, E. J. Chemical Modification Of Matrix Resin Networks with Engineering Thermoplastics.(1985).

Hildebrand, J.H. & Scott, R.L. the solubility of nonelectrolytes, 3rd Edition, Reinhold Pub. Co., New York (1950)

Hess, W, Vilgis, A T, Winter, H.H. Dynamical Critical-Behavior during Chemical Gelation and Vulcanization. (1988). *Macromolecules*, Vol. 21, No.8, pp.2536-2542

Hohenberg, C.P.; Halperin, I.B. Theory of Dynamic Critical Phenomena. (1977). *Rev. Mod. Phys.*, Vol.49, No.3, pp.435-479

Hourston,J.D.; Lane, M.J.; Macbeath,A.N. Toughening of epoxy resins with thermoplastics. Ii. Tetrafunctional epoxy resin-polyetherimide blends. (1991). Polymer International,Volum. 26, No 1, pp.17–21

Hwang, W.J; Cho,K; ParK, E.C.; Huh, W. Phase Separation Behavior of Cyanate Ester Resin/Polysulfone Blends.(1999). *J. Appl.Polym. Sc.*, Vol. 74, No.1, pp.33–45

Inoue, T. Reaction-induced Phase Decomposition in Polymer Blends. (1995). Progress Polymer Science, Vol.20, pp.119-153

Kim,B.S.; Chiba, T.; Inoue, T. A New Time-Temperature-Transformation Cure Diagram for Thermoset/Thermoplastic Blend: Tetrafunctional Epoxy/Poly(Ether Sulfone).(1993). *Polymer*, 1993, Vol. 34, No.13, pp. 2809-2815

Kim, M.; Kim, W.; Youngson, Choe, Y.S; Park, M.J.; Park, S.I. Characterization of Cure Reactions of Anhydride/ Epoxy/Polyetherimide Blends.(2002). *Polym. Int.*, Vol.51, No.12, pp.1353–1360

Kissinger, H.E. Reaction Kinetics in Differential Thermal Analysis. (1957). *Analytical Chemistry*, Vol.29, No.11, pp.1702-1706

Launey, E. M.; Buehler, J. M.; Ritchie, O. R. on the Mechanistic Origins of Toughness in Bone. (2010). *Annual reviews Materials Research*, Vol.40, p25-53

Lewin, L. J.; Maerzke, A.K; Nathan, E. Schultz, E.N.; Ross, B.R.; Siepmann, I.J. Prediction of Hildebrand Solubility Parameters of Acrylate and Methacrylate Monomers and Their Mixtures by Molecular Simulation.(2010). *J. Appl .Polym. Sci.*, Vol. 116, No.1, pp.1–9

Macosko, C. W.; Millerlb, D. R. A New Derivation of Average Molecular Weights of Nonlinear Polymers (1976). *Macromolecules*, Vol.9, No. 2, pp.199-206

Martuscelli, M; Musto,P.; Ragosta, G. 1996-10-01. *Advanced Routes for Polymer Toughening.* Elsevier Science Publishing Company, ISBN-10 / ASIN: 0444819606. Sara Burgerhartstraat 25, P.O. Box 211,1000 AE Asterdam, the Netherland

Mijovic, J.; Andjelic, S. Study of the mechanism and rate of bismaleimide cure by remote in-situ real time fiber optic near-infrared spectroscopy. (1996). *Macromolecules*, Vol.29, No.1, pp239-246

Montserrat, S.; Flaque, C.; Calafell M.; Andreu G.; Malek J. Influence of the Accelerator Concentration on the Curing Reaction of an Epoxy-Anhydride System. (1995).*Thermochimica Acta*, 269:213-229

Mours, M. & Winter, H.H. Time-Resolved Rheometry. (1994). Rheologica Acta, Vol.33, No.5,pp.385-397

Muthukumar, M. Screening Effect on Viscoelasticity Near the Gel Point. (1989). *Macromolecules*, Vol.22, No.12, pp.4656-4658

Oyanguren, A. P.; Aizpurua, B.; Gaamte, J. M.; Riccardi, C. C.; Cortazar, D. O.; Mondragon, I. Design of the Ultimate Behavior of Tetrafunctional Epoxies Modified with Polysulfone by Controlling Microstructure Development. (1999). *J.Polym. Sci., Part B: Polym. Phys.*, Vol. 37, No.19, pp.2711–2725

Oyanguren, A. P.; Galante, J.M.; Andromaque, K.; Frontini, M.P.; Williams, J.J.R. Development of bicontinuous morphologies in polysulfone–epoxy blends. (1999). Polymer, Vol.40, No.19,pp.5249–5255

Ohnaga, T.; Chen, W.J.; Inoue, T. Structure Development by Reaction-Induced Phase-Separation in Polymer Mixtures - Computer-Simulation of the Spinodal Decomposition under the Non-Isoquench Depth. (1994). *Polymer*, 35(17): pp3774-3781

Onuki,A. Late Stage Spinodal Decomposition In Polymer Mixtures.(1986). *J. Chern. Phys.*, Vol.85, No.2, pp.1122-1125

Onuki, A.; Nishimori, H. Anomalously Slow Domain Growth due to a Modulus Inhomogeneity in Phase-separating Alloys, (1991). *Phys. Rev. B*, Vol.43,No.16, pp 13649-13652

Onuki, A. ; Taniguchi,T. Viscoelastic Effects in Early Stage Phase Separation in Polymeric Systems.(1997). *J. Chem. Phys.* Vol. 106, No.13, pp.5791-5770.

Onuki, A. & Taniguchi, T. Viscoelastic Effects in Early Stage Phase Separation in Polymeric Systems.(1997). J. Chem. Phys., Vol.106, No.13, pp.5761-5770

Oono, Y.; Puri S. Study of Phase-Separation Dynamics by Use of Cell Dynamical Systems. I. Modeling.(1988). *Phys. Rev. A* Vol.38, No.1, pp.434–453

Pearson, R. A.;Yee, A.F. Toughening Mechanisms in Thermoplastic-Modified Epoxies: 1.Modification Using Poly(Phenylene Oxide). (1993). *Polymer*, Vol.34, No.17, pp.3658-3670

Ruccardi, C. C.; Borrajo, J.;Williams, J. J. R.; Cirard-Reydet, E.; Sauterau, H.; Pascault, P.J. Thermodynamic Analysis of the Phase Separation in Polyetherimide-Modified Epoxies.(1996). *J. Polym.Sci., Part B: Polym.Phys.*, Vol. 34, No.8, pp.349-356

Riccardi, C.C.; Bborrajo, J.; Meynie, L.; Fenoulillot, O.F.; Pascault, P.J. Thermodynamic Analysis of the Phase Separation during the Polymerization of a Thermoset System into a Thermoplastic Matrix. I. Effect of the Composition on the Cloud-Point Curves. (2004). *J. Polym. Sci., Part B: Polym. Phys.*, Vol. 42, No.8, pp.1351–1360

Riccardi C.C.; Bborrajo J.; Meynie L.; Fenoulillot O.F.; Pascault P.J. Thermodynamic Analysis of The Phase Separation During The Polymerization Of A Thermoset System Into A Thermoplastic Matrix. II. Prediction of The Phase Composition And The Volume Fraction Of The Dispersed Phase. (2004). *J. Polym. Sci., Part B: Polym. Phys.*, Vol. 42, No.8, pp.1361–1368

Riccardi, C.C.; Borrajo, J.;Williams, R.J.J. Thermodynamic Analysis of Phase-Separation in Rubber-Modified Thermosetting Polymers - Influence of the Reactive Polymer Polydispersity. (1994). *Polymer*, Vol.35, No.25, pp5541-5550

Ritzenthaler, S.; Court,F.; David, L.; Girard-Reydet, E.; Leibler, L.; Pascault, P.J. ABC Triblock Copolymers/Epoxy-Diamine Blends. 1. Keys To Achieve Nanostructured Thermosets.(2002). *Macromolecules*, Vol.35, No16, pp.6245-6254

Ritzenthaler, S.; Court, F.; David, L.; Girard-Reydet, E.; Leibler, L.; Pascault P.J. ABC Triblock Copolymers/Epoxy- Diamine Blends. 2. Parameters Controlling the Morphologies and Properties.(2003). *Macromolecules*, Vol.36, No.1, pp.118-126

Rozenberg, A.B.; Boiko, G.N.; Morgan, R.J.; Shin, E.E. the Cure J. Mechanism of the 4,4'-(N,N'- bismaleimide) diphenylmethane -2,2'-diallylbisphenol A System. (2001). *Polymer Science Series* A, Vol.43, No.4, pp.386-399

Ruseckaite, R.A.; Hu, L.; Riccardi, C.C.; Williams, R.J.J. Castor-Oil Modified Epoxy Resins as Model Systems of Rubber-Modified Thermosets. 2: Influence of Cure Conditions on Morphologies Generated. (1993).*Polym. Int.*, Vol.30, No.3, pp.287-295

Scanlan, J.C.; Winter, H. H. Composition Dependence of the Viscoelasticity of End-linked Poly(dimethylsiloxane) at the Gel Point.(1991). *Macromolecules*, Vol.24, No.1, pp47-54

Shibahara, S;Yamamoto, T.; Motoyoshiya, J.; Hayashi, S. Curing Reactions of Bismaleimido Diphenylmethane with Mono- or Di-Functional Allylphenols-High Resolution Solid-State C-13 NMR Study. (1998). *Polym. J.*, Vol.30, No.5, pp.410-413

Simon, S.L.; Gillham, J.K. Thermosetting Cure Diagram: Calculation and Application. (1994). *J. Appl. Polym. Sci.*, Vol.53, No.4, pp709-727

Tanaka, H. Viscoelastic Model of Phase Separation. (1997), *Physical Review E*. Vol.56, No.4, pp4451-4462

Taniguchi, T.& Onuki, A. Network Domain Structure in Viscoelastic Phase Separation. *Phys. Rev. Lett.*, Vol. 77, No.24, pp.4910-4913

Utracki L.A.; Simha R. Statistical Thermodynamics Predictions of the Solubility Parameter.(2004). *Polym. Int.*, Vol.53, No.3, pp.279–286

Williams, J. J. R.; Boris A. Rozenberg, A.B.; Pascault, P.J. Reaction-induced phase separation in modified thermosetting polymers. (1997). *Advances in Polymer Science, Polymer Analysis Polymer Physics*, Volm.128, pp.95-156

Wisanrakkit, G.&Gillham, J.K. The Glass Transition Temperature (Tg) as an Index of Chemical Conversion for a High-Tg Amine/Epoxy System: Chemical and Diffusion Controlled Reaction Kinetics. (1990). *J. Appl. Polym. Sci*, Vol.41, No11-12, pp.2885-2929

Xiong, Y.; Boey, F.Y. C.; Rath, K.S. Kinetic Study of the Curing Behavior of Bismaleimide Modified with Diallylbisphenol A. (2003). *J. Appl. Polym. Sci*, Vol.90, No.8, pp.2229-2240

Xu, J.J.; Holst, M.; Rullmann, M.;Wenzel, M.; Alig, I. Reaction-Induced Phase Separation In A Polysulfone-Modified Epoxy-Anhydride Thermoset. (2007). *J. Macromol. Sci. Phys.*, 46(1):155-181

Xu, Y.Z. & Zhang, X.J. Inversed Polarized Hotstage Microscope with High Resolution, Long Working Distance and High Temperature Duration. Chinese Patent. 2007200666499

Yi X.S. (2009). Research, Development and Enhancement of High-performance Polymer Matrix Composites for Aerospace in China.Proceeding of ICCM17, Edinburgh, UK, 27-31 July 2009

Yi X.S. Research and Development of Advanced Composite Materials Technology (in Chinese). National Defense Press, Beijing, 2006

Yi, X.S.; An, X.F.; Tang, B.M.; Pan, Y. Ex-Situ Formation of Periodic Interlayer Structure to Improve Significantly the Impact Damage Resistance of Carbon Laminates. (2003). *Advance Engineering Materials*, Vol.10, No.10, pp.729-732

Yi, X.S.; An, X.F. Developments of High-performance Composites by Innovative Ex situ Concept for Aerospace Application. (2008). *J. Thermoplast. Compos.*,Vol. 22, No.1, pp.29-49

Yi, X.S.; Xu, Y.H.; Cheng, Q.F.; An, X.F. Development of Studies on Polymer Matrix Aircraft Composite Materials Highly Toughened. (2008). *Science & Technology Review*, Vol.26,No.6, pp84-92 ISSN: 1000-7857 CN:11-1421/N

Yu Y.F.; Wang, M.H.; Gan, W.J.; Tao, Q.S.; Li, S.J. Cure Induced Viscoelastic Phase Separation in Polyethersulfone Modied Epoxy System.(2004).*Phys.Chem. E*, Vol.108,pp.6208-621

Zhang, H.D.Theoretical Studies and Computer Simulations of Phase Separation Kinetics of Polymer Mixtures. (1999). Ph. D. Dissertation, Fudan Unversity

Zhang, X.J. the Rheology and Morphology of Thermoplastic Modified Thermoseting Systems. (2008). Ph.D. Dissertation, Fudan University

Zhang, X.J.;Xu, Y.Z. *Proceeding of Advances in Rheology*,2006,138-144, Shandong University Press

Zhang, X.J.; Yi, X.S.; Xu, Y.Z. Cure Induced Phase Separation of Epoxy/DDS/PEK-C Composites and its Temperature Dependency. (2008). *J. App. Polym.Sci.*, Vol.109, No 4, pp2195-2206

Zhang, X.J.; Yi, X.S.; Xu, Y.Z. the Effect of Chemical structure on the Phase Separation Time/temperature Dependence in Thermoplastics Modified Thermosetting Systems.(2008). *Acta Polymeric Sinica*, No.6, pp583-591

Zhang, X.J.;Yi, X.S.; Xu, Y.Z. the Time/temperature Relationship during Phase Separation of Different Thermoplastic Modified Thermosetting Systems. (2007). *Acta Polymeric Sinica*, No 8, pp715-720

Zhang, X.J.; Yi, X. S.; Xu, Y.Z. Rheology and Morphology Development During Phase Separation and Gelation of Phenolphthalein Polyetherketon Modified Epoxy Resins. *Proceeding of the 22th Annual Meeting of the Polymer Processing Society*, pp527. Yamagata, Japan, Polymer Processing Society. July 2-6,2006

Zucchi, A.I.;Galante, J. M.; Borrajo, J.; Williams, J. J. R. A Model System for the Thermodynamic Analysis of Reaction-Induced Phase Separation: Solutions of Polystyrene in Bifunctional Epoxy/Amine Monomers. (2004). *Macromol. Chem. Phys.*, Vol.205, pp676-683

5

Processing of Carbon Fiber/PEI Composites Based on Aqueous Polymeric Suspension of Polyimide

Liliana Burakowski Nohara[1,2], Geraldo Maurício Cândido[2], Evandro Luís Nohara[3] and Mirabel Cerqueira Rezende[2]
[1]*State University of São Paulo - UNESP*
[2]*Institute of Aeronautics and Space, Department of Aerospace Science and Technology*
[3]*University of Taubaté, Unitau*
[1,2,3]*Brazil*

1. Introduction

Thermoplastic composites materials are being more frequently used in a wide range and variety of structures such as automotive and aerospace components. These applications often demand a unique combination of properties including high thermal and oxidative stability, toughness, solvent resistance, and low dielectric constant, etc (Nohara, 2005).

Thermoplastic have some distinct advantages over thermoset composites such as: high ductility and toughness, facility of processing and recycling potential. For applications that require high performance, the most commonly used thermoplastic are PEI (polyether imide), PPS (polyphenylene sulfide) and PEEK (polyether ether ketone) (Kong, 1996; Nohara, 2005; Nohara et al., 2005).

Thermoplastic can be classified by their molecular structure as either amorphous or semicrystalline polymer. Semicrystalline polymers crystallize to form structures with excellent chemical resistance, mechanical properties, and high service temperature, whereas the amorphous are unable to form ordered structures, showing up either in glassy or rubbery state. The glass transition it is not just determined by the chemical structure but, also, by the available free volume between the chain, allowing its rotational movement (Cogswell, 1992; Nohara, 2005).

The term amorphous implies that the polymeric chain it is present in a "entangled" without any degree of local order, unlike the semicrystalline polymer, that has a certain crystallographic orientation (Cogswell, 1992). One of the advantages of amorphous system is that they don't crystallize, since that it shows less a "variable" to be rated in the processing. A real advantage of amorphous polymer is that there is a slight variation in its volume between the solid and molten state, since there is no change in its specific mass associated with the formation of crystalline areas. Thus, such materials are less likely to be submitted to distortion in the cooling process, in the processing operation and for the composite material, it occurs a minor level of internal stress (Cogswell, 1992).

The aerospace industry has been researching more about the amorphous thermoplastic polymer and has found a meanful usage of they in aircrafts, especially where the high temperature performance is demanded (above 200°C), as well as a good resistance to solvents (Cogswell, 1992). Among the polymer high performance thermoplastic that has been used in aerospace industry, is the PEI, which is an amorphous polymer, transparent with amber colour that has been used since 1982.

The material is inherently flame resistant with low smoke emission, attending the established standards for the aircrafts interior and the requirements of the FAA (Federal Aviation Administration). It is soluble to solvents partially halogenated (such as the methylene chloride), but resistant to alcohol, acids and solvents made of hydrocarbon, bearing well the gama rays and ultraviolet radiation, characteristics of the semicrystalline polymer. The amorphous structure of the PEI contributes for the excellent dimensional stability and the mechanical proprieties highly isotropic, comparing to other semicrystalline (Offringa, 1996; Sroog, 1991).

Its high glassy transition temperature (217°C) and high modulus at high temperature are resulting from the rigid imide groups in their chemical structure. However, the ether groups justify the flexibility of the molecular chain. PEI also demonstrates good electrical properties and remains stable over a wide range of temperatures and frequencies. The high glass transition temperature allows the PEI to be used intermittently to 200°C. O PEI it stands out medical procedures, where the medical and dental equipments demand frequent sterilization. O PEI is also used for primary and secondary aircraft structure, radomes and stealth panels for military use and snow boards (Offringa, 1996; Jenkins, 1999).

The composite materials show different interfaces between its constituents which tend to govern the properties of material. This interface is defined as boundary surface between the reinforcement and the matrix and needs to be strong enough in order to allow stress transfer from the matrix to the reinforcement.

As the composite can be manufactured in high temperature and the reinforcement tends to have a lower CTE (coefficient of thermal expansion) than the matrix, the composite at room temperature tends to have residual stress, such that the reinforcement is under compression. The residual stress is particularly large when the reinforcement is in the form of fibers. The residual stress affects both the reinforcement and the interface between reinforcement and matrix (Chung, 2000).

However, besides the interface, the composite also have an interphase which is defined as the region formed by the interaction between the reinforcement and the matrix, possessing local properties different from those of the bulk matrix. The size and type of interphase varies strongly and depends on the nature of the fiber and its surface as well as on the polymer matrix. Because stress transfer in the composites should be influenced by the interphase, optimization of the mechanical properties of composites requires an extensive knowledge of the behaviour of the interphases and its effect on mechanical performance (Brodowsky et al., 2010; Pompe, Mäder, 2000; Reifsnider, 1994).

Interactions in the interphase region in thermoplastic composites depend on many factors such as matrix morphology, fiber surface conditions, presence of residual stresses, fiber and matrix elastic moduli as well as the presence of reactive functionalities. These functionalities can be obtained by coating the reinforcement or treating its surface with an interfacial bonding agent or coupling agent (Gao, Kim, 2001; Nohara et al., 2005).

Several methods have been developed to improve the fiber surface wettability or to increase the quantity of surface functional groups. The interfacial bond between the carbon filaments and the resin matrix can be enhanced by enlarging the surface area, which provides more points of contact/anchorage between the fiber and the matrix, or by enhancing the physicochemical interaction between the components (Burakowski, 2001; Nohara et al., 2005; Zielke, et al., 1996; Pittman Jr. et al., 1997).

Oxidation methods consist of oxidizing the carbon fiber in a liquid or gas environment to form oxygen-containing functional groups such as carboxyl, carbonyl, lactone and/or hydroxyl groups on the surface of the fiber, while simultaneously increasing the surface area of the carbon fiber. All these methods contribute to improve the stress transfer from the relatively weak and compliant matrix to the strong and stiff reinforcing fibers (Burakowski, 2001; Nohara et al., 2005; Zielke, et al., 1996; Pittman Jr. et al., 1997; Yue et al., 1999; Lee, Kang, 1997).

Another method to improve the adhesion between the reinforcement and matrix consist in applying a thin polymer layer (normally in lower than 1% of total composite mass), also known as coating or sizing on the reinforcement surface. This interphase area can also be developed spontaneously, by the matrix interactions with the carbon fiber surface (Chuang & Chu, 1990; Texier, A. et al, 1993).

The most used coating in carbon fibers are made of epoxy, phenolic and, furfurylic resins among others. However, these bonding agents don't bear temperature near to high performance thermoplastic matrix processing temperature, as the PEI, PPS, PEEK (>300 °C). In these case, can either be recommended the polyimide, due to the its high solvent resistance and high service temperature like the polymers above named (Burakowski, 2001; Nohara, 2005).

Polyimides (PIs) are a class of thermally stable high performance polymers that continue to gain importance in a wide variety of applications such as high temperature adhesives, microelectronics, membranes, photosensitive materials, matrix materials for composites and as reinforcement coating (Gao & Kim, 2001; Bessonov & Zubkov, 1993; Mittal, 1984; Ghosh & Mittal, 1996).

These applications are possible due to the many advantageous properties exhibited by polyimides, including excellent mechanical, thermal and adhesive properties, good radiation resistance, low dielectric constant, chemical resistances, there are no known organic solvent for the aromatic polyimides, and it has been used successfully in applications at temperatures as low as -269°C and as high as 400°C (Ghosh & Mittal, 1996; Saeed & Zhan, 2006; Srinivas et al., 1997).

In the case of composite applications, PIs have been successfully used as reinforcement coating between carbon fibers and high performance thermoplastic matrices such as PEEK, PEI and PPS (Chuang & Chu, 1990).

Recently, it has been possible to use PIs as interphase in structural thermoplastic composites by employing a new pre-impregnation and composite manufacture technique, namely, aqueous suspension prepregging (Texier et al., 1993; Gardner, 1998; Nohara, 2005). With this technique has been possible to improve the mechanical and chemical properties of the interface in thermoplastic composites.

The aqueous suspension prepregging method, was first proposed by Virginia Polytechnic Institute and State University researches and there is a few literature about this subject (Texier et al., 1993; Brink, Lin, Riffle, 1993; Gardner, 1998).

The aqueous suspension prepregging technique uses PAA (polyamic acid), a polyimide precursor, allowing the polymer matrix to be applied to the reinforcement together with the interphase forming polymer during a single pre-impregnation step (Brink, Lin & Riffle, 1993).

The aqueous suspension prepregging technique uses PAA neutralized with a base to produce the PAA salt. The polymer matrix (powder form) is then dispersed in an aqueous solution of the PAA salt, which acts as a dispersant and electrostatically stabilizes the suspension via the interaction with the surface of the matrix powder particles (Texier et al., 1993). Carbon fibers or fabrics are then coated with the polymeric suspension.

In a second phase of the processing, the PAA is thermally converted in a polyimide that can form an interphase region between the reinforcement and the polymer matrix, acting as coupling agent (Texier et al., 1993; Yu & Davis, 1993).

The purpose of this study is to compare two methods of processing thermoplastic composite: hot compression molding and aqueous suspension prepregging, showing that the latter method uses the insertion of a polyimide interface in composite. For polyimide interphase had been used 5 PAA/PI: BTDA/DHPr; BTDA/ODA; BTDA/DDS; BTDA/BisP e PMDA/ODA. As polymeric matrix in the composite processing, with carbon fiber, PEI was used. The PI synthesized as dust, they were characterized by the DSC technics (Differential scanning calorimetry), TGA (thermogravimetry), SEM (Scanning electron microscopy) and FTIR (Fourier transform infrared spectroscopy). The composites obtained in both manufature processes were compared and assessed by the technics of ILSS (interlaminar shear strength) and SEM.

2. Materials and methods

2.1 Reinforcement

As reinforcement plain weave carbon fiber was used, obtained from Hexcel Composites, with sizing compatible to epoxy resin. This fabric is composed of 3,000 filaments with a diameter of about 7 μm each. The carbon fiber fabric were cut in small square pieces (50 mm X 50 mm) and washed with acetone to eliminate the sizing.

2.1.1 Surface treatment of carbon fiber

Carbon fibers surface were treated, in order to verify the improvement of interface properties. This procedure was carried out in concentred nitric acid (Synth, 97%). Enough fabric to obtain a composite lot was treated (to acquire aqueous suspension prepregging based on BTDA/ODA).

2.2 Polymeric matrix

PEI, an amorphous thermoplastic polymer, was used. The PEI has T_g (glass transition temperature) of 217°C, is marketed by GE Plastics under trade name Ultem 1000, in pellets form and transparent amber color; its structure is shown in Figure 1.

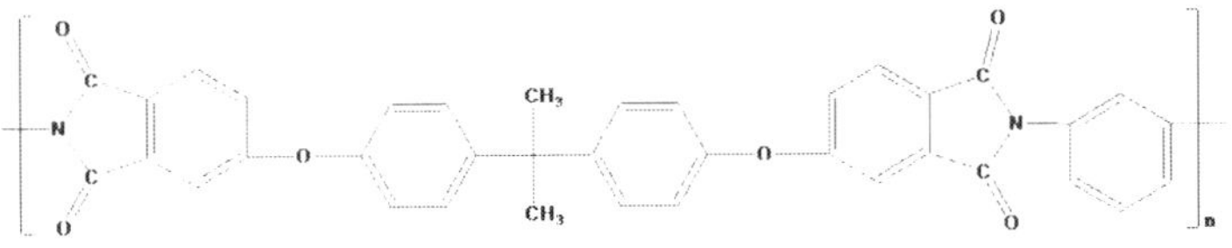

Fig. 1. Chemical structure of PEI.

2.2.1 Micronization of polymeric matrix

As the PEI was acquired commercially, in the form of pellets, it was necessary to micronize this material, so that it could be used to obtain polymeric suspension. Firstly, the PEI was grinded in a mill. For this, the pellets were immersed in N_2 liquid, for the purpose of fragilize them, in order to facilitate the milling. The grinded polymer in the mill presented a granulometry considered high (200 μm in average), to be used in the polymeric suspension technic, that is why it requires the micronization. For the micronization was used the Treu micronizer, the procedure was made three successive time.

2.2.1.1 Average size PEI measurement

The average size of the particles PEI was measured in a SEM from Zeiss, model 950 under variable pressure. The particles were sifted on a metallic specimen holder, which was covered by conductive carbon tape. The sample was blasted with compressed air to remove the excess of particles. The usage of SEM also played an important role, to observe the particles morphology after the micronization.

2.3 Polyimide interphase

2.3.1 Reagents

Dianhydrides: PMDA (pyromellitic dianhydride) and BTDA (3,3′,4,4′-benzophenonetetracarboxylic dianhydride). Diamines: DHPr (1,3-diamino-2-propyl alcohol); ODA (4,4′-diaminodiphenyl ether); DDS (4,4′-diaminodiphenyl sulfone) and BisP (4,4′-[1,4-phenylenebis(1-methylethylidene)]). Solvent: Acetone. All chemicals were purchased from Sigma Aldrich (97–99% purity) and the solvent was Merck.

2.3.2 Synthesis

The synthesis of the PI precursor, o PAA, was made according to literature (Asao, 2001; Nohara et al., 2007). Two solutions were prepared from equimolar quantities of the dianhydride and diamine in acetone (dianhydride or diamine/acetone = 0.004 mol/100mL). Both solutions were mixed at room temperature in a buchner flask at N_2 atmosphere, under ultrasonic agitation (40 kHz) for 4 hours. Monofunctional end-capper phthalic anhydride (from Vetec) was added to solution (16.10^{-5} mol). The end-capper has the function to limit the chain growth of PAA/PI which in smaller size, as an oligomer, tends to migrate to the fiber/matrix interface, acting as an interface/interphase creator on the bonding between the components of the composite. The conclusion of the reaction was monitored by thin layer chromatography. The chemical structures of PAAs/PIs are showed on the table below (Nohara et al., 2007).

PAA Structure	PI Structure	Polyimide Designation
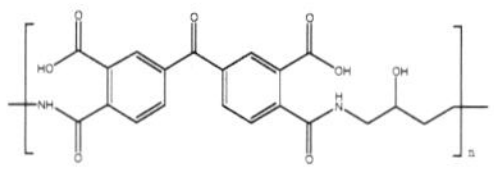	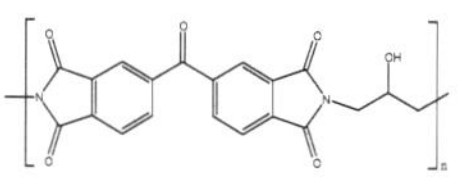	BTDA/DHPr
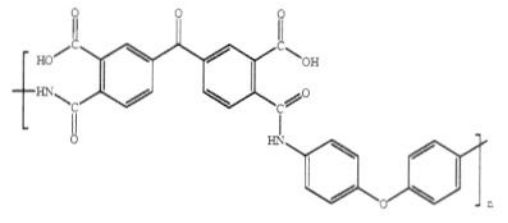	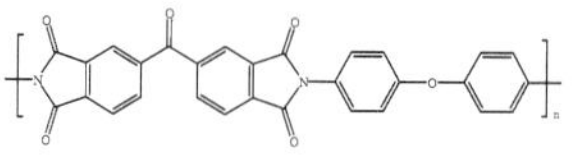	BTDA/ODA
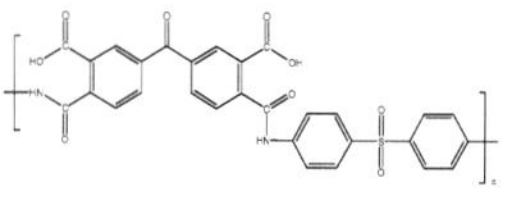	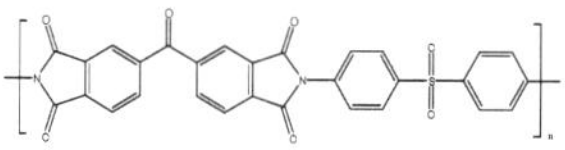	BTDA/DDS
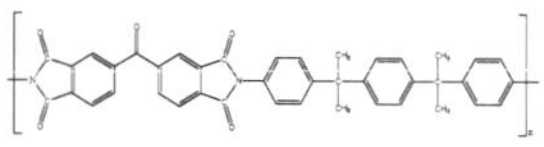	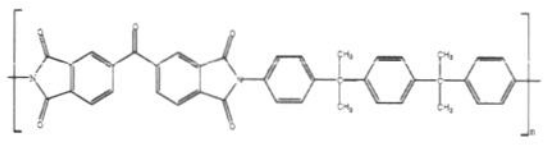	BTDA/Bis-P
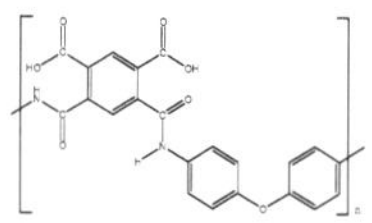	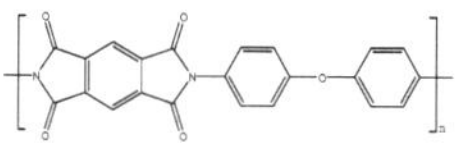	PMDA/ODA

Table 1. Chemical structures of PAA/PI.

2.3.2.1 Characterization of PI

The glass transition temperature from each PI was determined by DSC Pyris 1 Perkin Elmer under N_2 purge (20 mL/min). From each sample, approximately 13mg, was firstly imidized in the equipment and submitted to the cycle of : 1h-100ºC, 1h-200ºC and 1h-300ºC at 10ºC/min. After, the samples were cooled at ~100°C/min up to 150°C and submitted to a new cycle up to 550ºC (20ºC/min).

The thermal stability of the PIs was analyzed by the TGA. The sample were imidized in the equipment, following the same procedure applied on the DSC analysis. After this, the samples were cooled at room temperature. Subsequently, imidized samples were analyzed from room temperature up to 1000ºC (10ºC/min) to obtain T_d.

Fourier transform infrared spectroscopy (FT-IR) was used to confirm the presence of functional groups within both PAAs and PIs (BTDA/DHPr, BTDA/ODA, BTDA/DDS, BTDA/BisP and PMDA/ODA). Measurements were carried out on a FT-IR spectrometer Spectrum 2000 from Perkin Elmer, using the potassium bromide disk technique (1:400 mg).

2.4 Composites preparation

2.4.1 Hot compression molding

The preparation of composites using hot compression molding consist in spread the powdered polymeric over the carbon fiber fabrics, as showed on the Figure 2. In a mold, eighteen layers of carbon fiber fabrics (cleaned with acetone) were used intercalated with PEI micronized in an aluminum mold, in the ratio reinforcement:matrix of 50/50 (v/v) (Nohara et al., 2007).

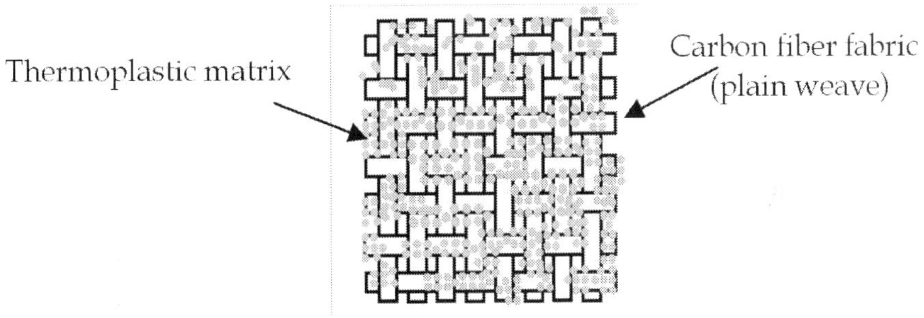

Fig. 2. Schematic diagram showing the carbon fiber/matrix composite assembly by hot compression molding.

2.4.2 Aqueous suspension prepregging

2.4.2.1 Preparation of poly(amic acid) salt

For the composites manufacture through aqueous suspension prepregging technic, were used PAAs (showed on the Table 1) as dispersants, with purpose of induce the matrix suspension of the polymeric particles (PEI) in solution PAA salt (obtained by the solubilization of PAA in NH_4OH) and so, obtain an effective wettability of carbon fiber fabrics by the system. In addition, the PAA played a very important role in the composites processed by this method: the interphase between the reinforcement and the matrix.

This method of composites processing it had already been studied previously by other scientist (Texier et al., 1993; Gardner, 1998). The last scientist mentioned in his work, he used three different bases: ammonium hydroxide (NH_4OH); tetramethyl ammonium hydroxide – (TMAH - ($N(CH_3)_4OH$)) and tripropylamine hydroxide (TPA - ($NH((CH_2CH_2CH_3)_3OH)$)) to obtain the composites based in PPS (poly(phenylene sulfide)) and PEEK (poly(ether ether ketone)) and he studied the influence of these three bases in the molar mass distribution, in the thermal properties and viscosity of interphase formed by Ultem and BTDA/Bis-P type polyimide interphase. Gardner noticed that, the bigger the size of the counterion ($NH_4^+ <$

$TMA^+ < TPA^+$) the better is the thermal properties interphase PI, bigger is the molar mass distribution and smaller is the viscosity molten. These results confirm that final properties of the interphase can be controlled and tailored according to the requirements and the usage of the composites.

In the current article, was choosen to use a single base or counterion: NH_4OH, since the main objective of this study is to identify the affinity of the PEI with the 5 PIs synthesized. With the purpose to study the effects of pH in the salt solutions of PAA in polymeric suspension of PEI, seven different solutions of NH_4OH were patterned, in the following pH: 11.70; 11.80; 11.90; 12.00; 12.10; 12.20, and 12.30.

It was used NH_4OH from the Synth (28-30%) and deionized water was added to the NH_4OH, to correct pHs. Then, salt solutions were prepared from PAA (NH_4PAA), from solubilization of each PAA in solution NH_4OH from pHs 11.70 to 12.30. The NH_4OH was used with stoichiometric ratio of 1.25:1 for the functionalities of PAA (two per unit). The 25% of stoichiometric excess of the base were used to assure the neutralization of the functionalities of amic acid to and keep the salt stability of polyamic acid, as studied by Texier and Gardner (Texier et al., 1993; Gardner, 1998).

2.4.2.2 Preparation of the polymeric suspension

After obtaining NH_4PAA, each beaker with salt was covered with a polymer film and heated at 60°C for 30 minutes, under magnetic stirring. Then, the solution was cooled up to room temperature. After this phase, the micronized PEI was added to NH_4PAA solution, and then homogenized with a magnetic stirrer for 30 minutes (Texier et al., 1993; Gardner, 1998; Nohara et al., 2007). The NH4PAA/PEI mass ratio was 0.05 and the concentration of PEI in the suspension was 10% mass (Gardner, 1998).

2.4.2.3 Preparation of the laminate

The carbon fiber fabrics were immersed in a beaker with polymeric suspension under stirring. The fabrics impregnated with NH4PAA were kept in room temperature for 24 hours. Thereupon, were taken to an oven under vacuum, were they preimidized under the following cycle: 1h-100°C; 1h-200°C e 1h-250°C a 10°C/min (the complete imidization of the PAA, occurred together the polymer matrix, in the consolidation of the composites). Next, the impregnated fabrics, were placed (in 18 layers), into a mold and were taken to a hydraulic hot press for the consolidation, that started heating from room temperature (~30°C) up to 310°C (10°C/min). It was established a ramp of 1h at 310°C, and applied in a pressure of 9.8 MPa for 30 minutes in this temperature.

The same procedure (consolidation) was applied to obtain the composites via hot compression molding.

2.4.3 Characterization of the composites

2.4.3.1 Interlaminar Shear Strength (ILSS)

The ILSS test was conducted according to ASTM D 2344. The test was made using 10 specimens obtained for each composites type, through the methods by the hot compression molding and aqueous suspension prepregging. The values of ILSS were calculated by the use of equation 1. The test was conducted by a test machine from universal Instron, in a test speed of 1.3 mm/min and a load cell of 5 tons.

$$F^{sbs} = 0.75 \times \frac{P_m}{b \times h} \tag{1}$$

where: F^{sbs} = is the short-beam strength (MPa)

P_m = is the maximum load during the test (N);

b = is specimen width (mm);

h = is the thickness (mm).

2.4.3.2 Scanning electron microscopy

The microscopic analysis of the composites surface fracture was made from sample submitted to interlaminar test shear strength. The surfaces for analysis by SEM were generated through a process that consisted in loading in shear, under very low speed until material rupture by delamination. Thereafter, the specimens generated during the loading were immersed in an isopropyl alcohol solution and distilled water (50:50 v/v) in a ultrasonic bath for 5 minutes. Then the specimens were blasted with compressed air for drying and were placed in a suitable holder. Carbon tapes and conductive silver ink were used to the electrical contact with the microscope. These analyses were accomplished in SEM Zeiss 950 and the results were registered with magnifications of 1000 and 2000 times.

3. Results and discussion

3.1 Reinforcement

3.1.1 Surface treatment of carbon fiber

The chemical attack of the carbon fiber, was accomplished to promote the surface changes, by a reliable introduction of the chemical polar groups and for the roughness rise (Nohara et al., 2005).

In the chemical treatment, carbon fiber fabrics (as received) were treated with a concentred nitric acid solution for 10 minutes at 103°C, with heating plate without agitation. The exposition time was chosen based on studies of Nohara et al. (Burakowski, 2001; Nohara et al., 2005) who concluded that the fibers treated superficially with nitric acid shown an increase roughness (measured by Atomic force microscopy and SEM) and a small decrease in tensile strength of single filaments of the carbon fiber (2143 ± 471MPa to 1924 ± 658MPa) - ASTM D-3379, due to removal of surface layers weakly attached to the fiber surface and the reduction of critical failures, that act as stress concentration. Analysis by XPS (X-ray photoelectron spectroscopy) it was also used and found the introduction of polar groups such as C-OH, C=O and COOH at the fiber surface treated with nitric acid (Burakowski, 2001; Nohara, et al., 2005).

After the chemical treatment, the fibers were thoroughly washed with freshly boiled deionized water and dried in an oven at 105°C for 2 hours. Carbon fiber fabrics treated were immediately kept in a desiccator under vacuum until its usage in composites preparation. Just one family was chosen for the research of carbon fiber fabrics with changed surface (carbon fiber-BTDA/ODA-PEI).

3.2 Polymeric matrix

3.2.1 Micronization of polymeric matrix

The polymeric matrix of PEI, after being grinded, passed through three micronizations consecutive and it was observed that this procedure led to the breakage of the particles, as expected and the smaller the particles size the more accentuated became the phenomenon of attraction among them and consequently, the packing. The micronized particles were separated by sieve 200–400 mesh coupled to a sieving machine with ultrasonic agitation. Just the obtained part in sieve 400 mesh was used in aqueous suspension prepregging.

3.2.1.1 Average size PEI measurement

The PEI particles, analyzed by MEV (Figures 3-a and -b) showed a rounded grain, like a cabbage, composed by many foils. This morphology of thin "foils" it was obtained probably due to the successive micronization suffered by the particles that made them to stick to one another, at the milling, they originated a packing of polymer layers. The PEI, as it can be observed on Figures 3-a and -b, showed after the micronization, the majority of its particles with sizes over than 30 μm.

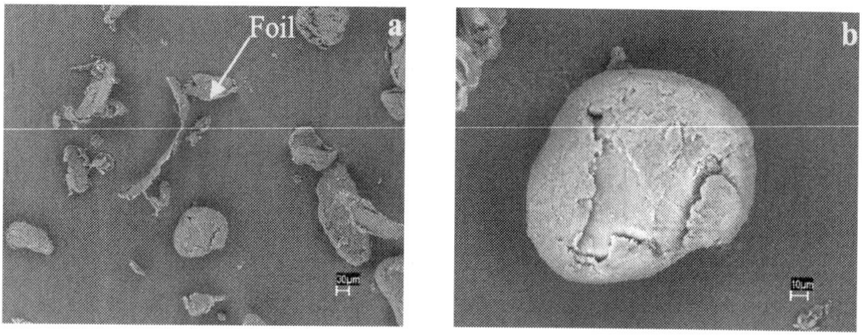

Fig. 3. Particle of PEI - a) 120X and b) 1000X.

3.3 Polyimide interphase

The synthesized PAAs showed the following colors: BTDA/DHPr = pale yellow; BTDA/ODA = yellow; BTDA/DDS = orange; BTDA/BisP = bright yellow and PMDA/ODA = golden yellow. Next, will be presented the analyses made by DSC, TGA and FT-IR of PAA and PI.

3.3.1 Characterization of PI

The measured glass transition temperature, Tg, of PI: BTDA/DHPr, BTDA/ODA, BTDA/DDS, BTDA/BisP, PMDA/ODA were measured by DSC, as showed on the table 2. T_g obtained for the PIs showed a slight difference from the one found in the literature, due to the changes in variable of the synthesis, such as: different solvents, methods and cycles of imidization, among others. There is a very interesting case to be emphasized, is the PI PMDA/ODA from literature (Sroog, 1991), which indicates a value of T_g = 399°C and a temperature of theoretical melting temperature of 595°C. However, as described in the same literature, at 510°C already occurs a mass loss of ~5%, which probably compromise the fusion identification by the overlap of the thermal events. The values found here, for this polyimide were different (T_g=437°C e T_d=498°C) due to a likely difference of the solvents and synthesis method used.

For the TGA analysis, the samples were imidized in the equipment, and it was checked, during the imidization cycle, that each one of them lost 21% of mass, concerning to the loss of solvents and to the water release from the synthesis and imidization. Next, the samples were cooled up to room temperature, were submitted to a second scanning, so that the thermal stability could be evaluated from each one of them (Table 2). The TGA analysis of the PIs synthesized show a beginning of decomposition temperature almost the same the found in the literature. The PI from the type BTDA/BisP, for example, shows a beginning of T_d decomposition temperature of 401°C (Table 2). Gardner (Gardner, 1998) explains the acquirement of this PI com T_d at about 489°C. The biggest difference found between the T_d it is due to a likely difference of the solvents and synthesis used: Gardner (Gardner, 1998) used NMP (n-methyl-2-pyrrolidone), a solvent of high boiling point solvent (202°C), while the current study was used acetone (56.5°C), with the purpose to obtain PI directly in powder form.

Polyimide	T_g (°C)		T_d (°C)	
		Reference		Reference
BTDA/DHPr	213	201*	310	310*
BTDA/ODA	305	329*	422	532*
BTDA/DDS	300	307**	396	400**
BTDA/Bis-P	248	270***	401	489***
PMDA/ODA	437	400**	498	505**

*Asao & Saito, 2001; **Sroog, 1991; ***Gardner, 1998.

Table 2. Glass transition temperature (T_g) and decomposition temperature (T_d) for the PI of the types: BTDA/DHPr, BTDA/ODA, BTDA/DDS, BTDA/BisP, PMDA/ODA.

The FT-IR analysis was used for the identification of synthesis of PAA or from the possible incomplete reaction found from the used reagents - which is characterized, mainly by carboxylic acid remaining group of dianhydride - and PI. PAA samples show similar absorption spectra in Figure 4-a, wherein bands near 1719-1740 cm^{-1} were assigned to the vibrational modes of acid groups for all samples. Bands near 1660 and 1537-1542 cm^{-1} were assigned to the vibrational modes of amide groups.

The bands related to acid and amide groups disappeared after imidization (Figure 4-b). The characteristic absorption bands of imide groups near 1780, 1722-1724, 1380 and 721-725 cm^{-1}, were present in the spectra (Figure 4-b). These absorption bands confirm the presence of imide groups. Table 3 presents the Infrared absorption bands of amides, imides and related compounds.

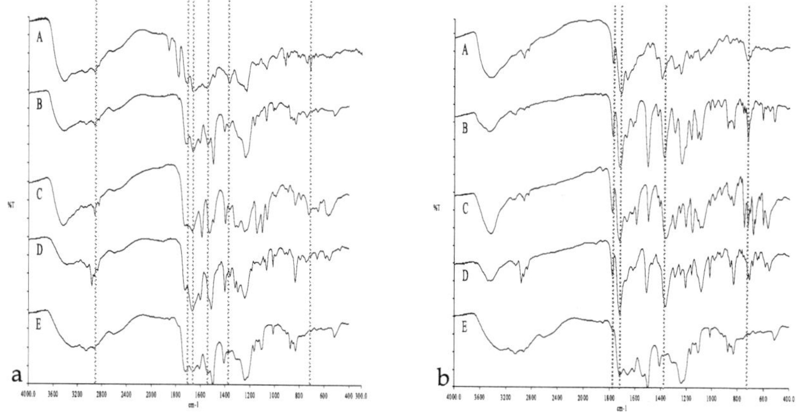

Fig. 4. FTIR spectrum of: a) PAA and b) PI: A) BTDA/DHPr, B) BTDA/ODA, C) BTDA/DDS, D) BTDA/BisP e E) PMDA/ODA.

	Absorption band (cm^{-1})	Origin
Aromatic imides	1780	C=O asymmetric stretch
	1720	C=O symmetric stretch
	1380	C-N stretch
	725	C=O bending
Amic acids	2900-3300	COOH and NH_2
	1710	C=O (COOH)
	1660	C=O (CONH)
	1550	C-NH
Anhydrides	1820	C=O
	1780	C=O
	720	C=O

Table 3. Infrared absorption bands of imides and related compounds.

3.4 Preparation of the composites

3.4.1 Hot compression molding

The obtained composites from molding compression were retired of the mold and cutted according to the test specification of ILSS.

3.4.2 Aqueous suspension prepregging

3.4.2.1 Preparation of the polymeric suspension

For the composites manufacture, through the aqueous suspension prepregging, 5 different PAAs were used as dispersant, with the purpose of causing suspension of polymeric matrix particles of the PEI in the salt solution in the PAA thus, obtaining a effectively wettability of the carbon fiber fabric by the system.

3.4.2.1.1 Investigation of the polymeric suspension behavior due to the use of pH different

The polymeric suspensions based in PEI were evaluated in relation to interaction with the 5 PAAs, under seven different values of pH: 11.70; 11.80; 11.90; 12.00; 12.10; 12.20; 12.30. The Figure 5 shows the polymeric suspensions of the PEI, prepared from the PAA type BTDA/ODA, which is representative for all the PAAs studied.

Observing the obtained suspensions from the PAA type BTDA/ODA, there was not found any evidence, related to the material suspension quantity - in the system there after the immediate agitation. Can also be observed that, the tonality of the seven solutions, under seven different pH did not suffer any color alteration. Despite of this, the original color of PAA-BTDA/ODA, obtained from the synthesis, is bright yellow, while the one from PEI it is bright amber, indicating that all the systems, suffered the same tonality changing. It was also observed the rise of the pH value, did not influence the NH_4PAA-PEI interaction and quantity of polymer in suspension. The same procedure was applied in 4 different PAAs and no meanful change was found. Thus, it was chosen the usage of pH 12 to obtain all composites by aqueous suspension prepregging.

Fig. 5. Polymeric suspension of NH_4PAA-PEI type BTDA/ODA.

3.4.2.1.2 Evaluation of the adhesion PEI/interphase/reinforcement

The Figures 6 a–b show the carbon fiber fabrics that were impregnated with the polymeric suspension of the NH_4PAA-PEI from the types: BTDA/ODA e PMDA/ODA, respectively, after imidization. Can be observed on the Figure 6-a there was a significant adherence of the polymeric matrix PEI to the carbon fiber fabrics, by the usage of the PAA/PI interphase type BTDA/ODA; since, placing the semipreg in vertical position no particle was displaced.

It was also identified that, pre-impregnation allowed the handling of the semipreg, by the perfect adherence of the particles, besides the perfect distribution of these particles. The Figure 6-b shows the aqueous suspension prepregging process provides a formation of a polymeric film stuck to the carbon fiber, easy to be seen by the strong adhesion of PEI particles each other, by the usage of PAA/PI type PMDA/ODA. Can also be observed that, polymeric film part can be unstuck and handling with the spatula. So, can be presumed that this system offers a perfect compatibility between PAA-BTDA/ODA or PMDA/ODA and PEI.

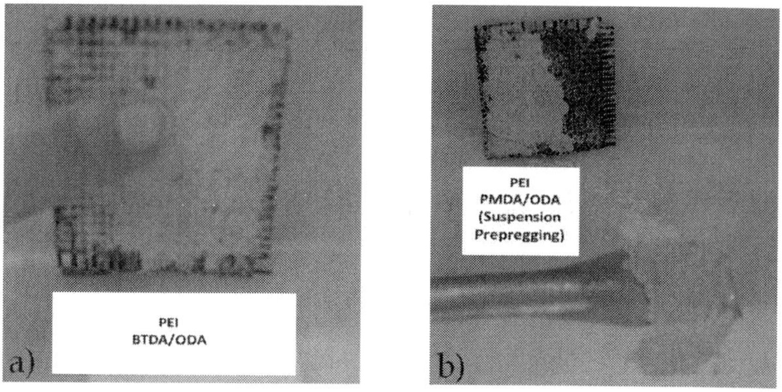

Fig. 6. The semipregs sample pre-imidized, obtained through a pre-impregnation via polymeric suspension - a) PAA-PEI type BTDA/ODA e b) PAA-PEI type PMDA/ODA.

3.4.2.1.3 Composites preparation – Hot compression molding X aqueous suspension prepregging

The main objective of the current study is to compare the acquirement process of composites by the hot compression molding and aqueous suspension prepregging. Once, interface/interphase region obtained by two processing methods can have a meanful difference and, considering that this region is one the composites components with meanful importance, all the variables must be considered.

At first, the obtained composites by aqueous suspension prepregging have a differential related with obtained composites by hot compression molding method: the PI interface/interphase. It is expected that the mechanical properties - that in the current study are evaluated by interlaminar shear strength - have an increase in values, using the method that makes use PI interphase.

However, as can be seen, some information related to both processing methods can also be obtained, by the simple observation of the contact between the carbon fiber fabric and the polymeric matrix, as shown on Figures 7-a and -b which show (1) fabric impregnated with the polymeric suspension of PEI-PMDA/ODA and (2) the covered fabric with the polymer PEI powder.

The Figure 7-a (2) shows that when fabric "2" (would be the hot compression molding method) it stands on horizontal position, all the polymer remains spread on the reinforcement. However, when the fabric is exposed to the vertical position (Figure 7-b(2)), can be noticed that the polymer matrix loose itself easily from the fiber fabric. It happens because the lamination process, for the composites acquirement by the hot compression molding method, it consist on spread the polymeric matrix in the reinforcement; while, the pre-impregnation method (Figure 7-a (1) e 7-b (1)) consist in use of an interphase. In this case, it favors adhesion between the polymeric matrix and the reinforcement, like an adhesive type. It allows the best spreading of polymer matrix on the fabrics, besides to favour the adherence and wettability before the composites consolidation.

The Figures 7-a (2) e -b (2) also show that the method that consist on spread the polymeric matrix on the fabric has one more interference: spreading the powder over the fabric, it enters easily through of weft and warp, not allowing a uniform spread of the polymer matrix on the fabric layer, by introducing a different quantities of matrix, of a layer to another of composite.

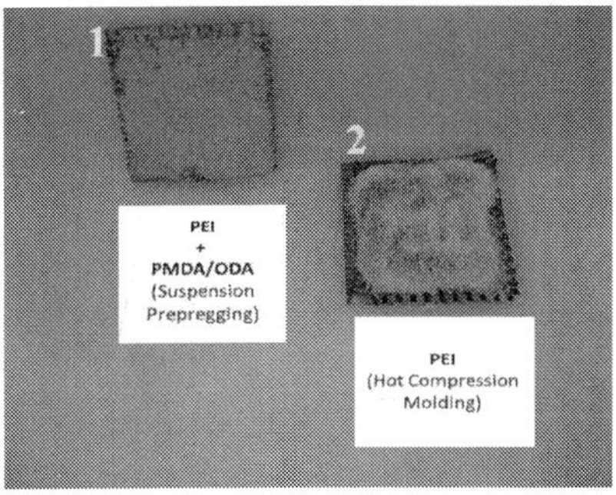

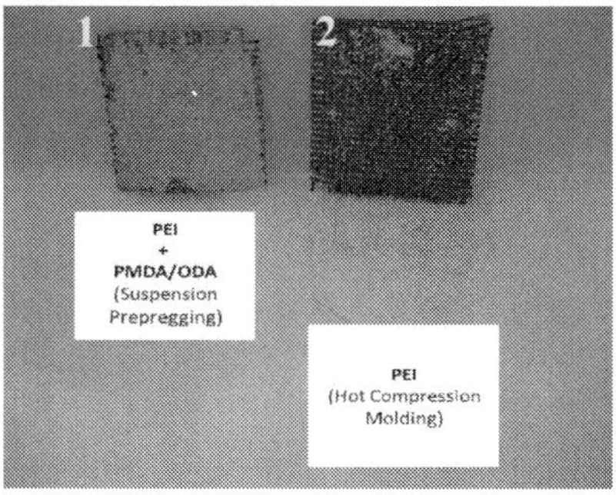

Fig. 7. (1) Impregnated fabrics with polymeric suspension PEI-PAA type PMDA/ODA e (2) PEI was spreaded on the carbon fiber fabric - hot compression molding.

3.4.3 Characterization of the composites

3.4.3.1 Interlaminar shear strength (ILSS)

The Table 4 relates the ILSS values and the respective standard deviations from the 7 lots of composites based in PEI, obtained by hot compression molding and aqueous suspension prepregging.

Interlaminar shear strength (ILSS) (MPa)	
Hot compression molding	
Carbono fiber/PEI composite	66.1 ± 5
Aqueous suspension prepregging	
PI-BTDA/DHPr	40.6 ± 3
PI-BTDA/ODA	52.5 ± 2
PI-BTDA/ODA/ fabric treated by HNO_3	80.6 ± 8
PI-BTDA/DDS	57.4 ± 5
PI-BTDA/Bis-P	74.7 ± 6
PI-PMDA/ODA	78.1 ± 7

Table 4. Interlaminar shear strength (ILSS) values of the thermoplastic composites manufactured by the hot compression molding method and aqueous suspension prepregging.

The analysis of curves (stress x displacement) showed, for all the composites, the occurrence of failure not catastrophic, in other words, didn't happen an abrupt break of the sample, after reach the maximum stress value. The composites showed interlaminar fracture, i.e. the crack formation in horizontal way.

The observation of Table 4 shows, in general way, that the obtained composites by the aqueous suspension prepregging method presented an improvement in the interlaminar shear strength values compared to the prepared composites using the hot compression molding. This is due to the usage of PAA/PI, besides acting as dispersant in the polymeric suspension, they also had an important role for the composites: as interphase creators.

The PI usage also favored the interdiffusion possibility of the PI into polymeric matrix, which resulted in a gradient of properties in the matrix, as it was evidenced by different ILSS values found, due to different PI polymeric chains.

The interphase material can have best and differentiated properties from the ones found in the matrix bulk. This is exactly the purpose of using the aqueous suspension prepregging in the current study: improve the interface/interphase. Thus, the 7 lots of composites were analyzed.

The PEI composites prepared by hot compression molding process, presented higher ILSS values, comparing from the ones prepared with aqueous suspension prepregging, just for the samples obtained PI-BTDA/DHPr interface. The decrease in ILSS values for the processed composites with this interphase, comparing to the hot compression molding process, it is related, probably, to the chemical structure of the PI. This is the only PI, among the 5 used in this study, that present less aromaticity, due to the structure of diamine (linear - DHPr).

Observing the lots of sample, it was proved that the prepared composites with the polymeric suspension based in PI-PMDA/ODA and BTDA/ODA/treated fabrics with HNO_3, they had better results of ILSS (18.1 and 21.9% respectively), comparing to the composites obtained by hot compression molding.

The difference of the ILSS values of the composites with different PI interphases, can be explained by a slight difference of polymeric chains. The PEI it is a polymer from the polyimides class and has part of its chemical structure similar to the PIs studied, with p-phenylene and imide groups that provide the stiffness to a part of the structure. PI BTDA/ODA, however, it is considered a flexible chain, due to the carbonyl group "bridges", between the benzene rings of the dianhydride and the ether between the benzene rings of diamines. This flexibility allows a better interfacial contact with the carbon fiber, resulting in stronger bond strength. In fact, in the last case, it is important to observe that there is a difference meanful between the ILSS values found for the composites that used PI-BTDA/ODA, related to the composite obtained with PI interphase equal, but which made the use of the treated reinforcement with HNO_3. The meanful increase of the ILSS property for this composite, comparing to the prepared composite with the same interphase, it is due to the probable interdiffusion chains mechanism of the PI into PEI, because when it is in contact with the treated fibers (now with high quantity of polar group and rugosity - as shown on the item 3.1.1) they were able to favour a strong chemical bond and mechanical anchorage.

It is important to emphasize, by using the aqueous suspension prepregging the PAA/PI is present as a oligomer in the system thus, it shows a tendency to the migrate to the interface between matrix and reinforcement. This facilitates the fiber wettability by PAA/PI, changing the interfacial resistance with the matrix.

The PMDA/ODA, unlike the BTDA/ODA, is PI the more rigid among the studied (it has 4 carbonyl groups attached to the same benzene ring in a coplanar conformation) it would be expected that had a smaller interaction with the carbon fiber. However, the difference in ILSS values of this interphase comparing to the others (BTDA/DDS, BTDA/BisP), it might be related to the interaction of part of its molecule with PEI, once these molecular structures can be overlapped, due to the conformation of its molecules can lead to points of attraction between chains and thus providing a strong interaction.

3.4.3.2 Electronic scanning microscopy

After the interlaminar shear strength test, the samples were analyzed, with respect the morphology of their fracture surfaces by SEM.

The samples for the analysis were obtained by loading in shear, with a very low speed, until the final charge for the production of a fracture surface by delamination. In overall, different morphologies were observed from 7 different manufactured composites (Figures 8 a-g).

The figure 8-a (FC/PEI - hot compression molding) shows the micrography of fracture surface of the laminate obtained by loading to failure. This micrography shows that there was disruption of the laminate in the fiber direction and that there is the presence of channels and ridges on the surface of fracture, which may have been occasioned by the forced loading. Can be observed (red arrow) which the outline of most of the carbon fiber filaments; it is covered with a thin PEI layer, suggesting that there was a wettability of reinforcement by the matrix, but there are also discovered fibers.

The Fig. 8-b shows a fracture surface (caused by the loading) from the obtained composites with a PI-BTDA/DHPr interphase. The micrography shows that there are carbon fiber filaments discovered, probably due to the effort of loading. In this case, seems to have a slight excess matrix among the fiber filaments, it maybe might contribute for the low ILSS values obtained. It can also be observed that the polymer presents aspect more dry and fragile, related to the carbon fiber sample/PEI without the PI interphase (Fig. 8-a) which looks have behavior more ductile.

The Fig. 8-c shows (interphase BTDA/ODA), the presence of uncovered fibers free from polymeric matrix on the laminated surface and also matrix excess in other regions. These observations don't match with ILSS values found in this composites lot. What could have happened is that the region was selected for microscopy observation wasn't representative for the composite as a whole.

The Fig. 8-d shows a composite obtained from the same interphase (BTDA/ODA) but, was used carbon fiber fabrics treated with HNO_3. In this case, are observed regions where the polymeric matrix was removed (red arrow) with the interphase, from the loading to the rupture. It was also observed, a differentiated texture from the fracture surface, with respect to found in other composites, suggesting that there was good wettability of reinforcement by the matrix and consequently, a strong adherence, which was confirmed by the interlaminar shear strength test.

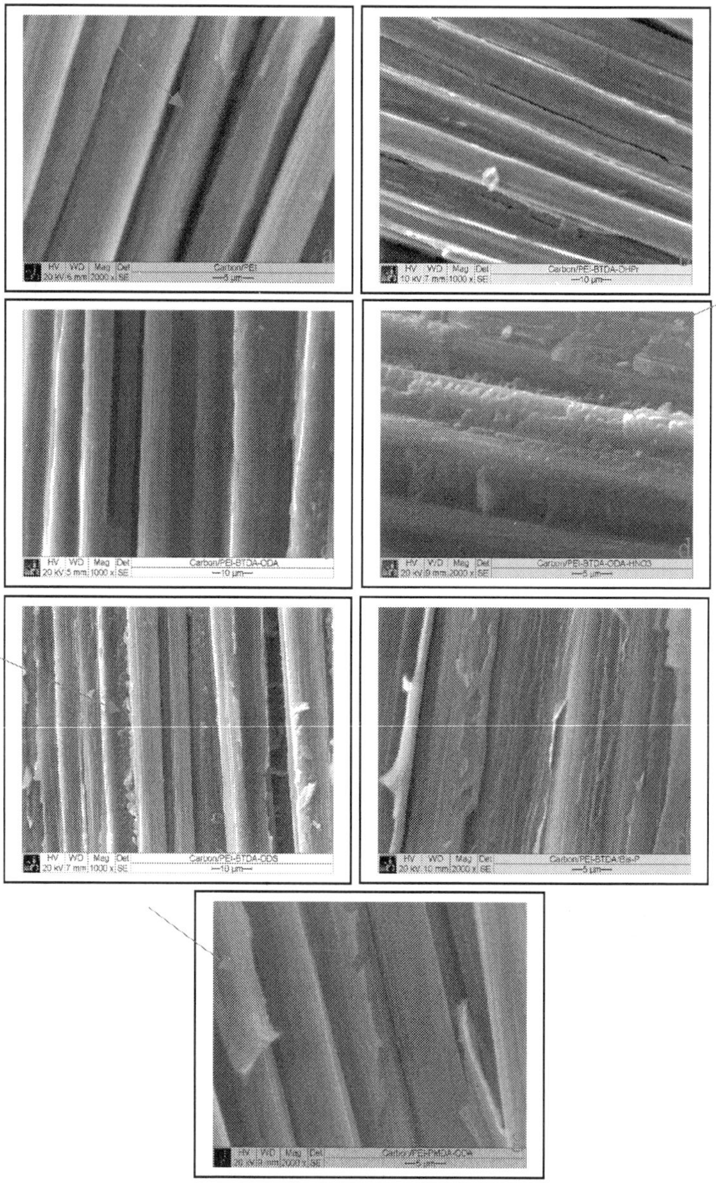

Fig. 8. The fracture surface of the manufactured composites with: a) the hot compression molding and aqueous suspension prepregging b) PI-BTDA/DHPr; c) PI-BTDA/ODA; d) PI-BTDA/ODA-treated fiber with HNO_3; e) PI-BTDA/DDS; f) PI-BTDA/BisP and, g) PI-PMDA/ODA.

The Fig. 8-e shows a micrography obtained from the composite of interphase BTDA/DDS. Can also be observed that there was pull out of carbon fiber filament (caused by loading) which left a gap among them (channels). Cleaned carbon fiber were found, that was not wet by the PEI with the interphase BTDA/DDS and in some places, matrix excess (red arrow). These results match with the ILSS test, as the values found in this lot were very low (lower than the ones found for the obtained composites with hot compression molding).

The Fig. 8-f shows a composites made using the PI-BTDA/BisP interphase. It Can be observed that there was not any failure on interface fiber/matrix. There are places where the polymer was pulled out, leaving the uncovered fiber, but can also be observed that, the polymer covered the fibers in many layers. The fiber wettability with system PI-BTDA/BisP-PEI interface seemed to be satisfactory.

The Fig. 8-g shows uncovered fibers, but suggesting that were covered by the system PMDA/ODA performed prior to loading to generate the fracture surface for analysis by SEM. The micrography shows clearly that PEI shows a ductile behavior (red arrow). This situation could have been generated at the loading moment; which may have introduced a certain temperature the composites by the force applied to the material.

For all these cases it was not possible to check a bondary between PI and the polymer matrix PEI, indicating that there was an interdiffusion between both polymers.

4. Conclusion

In this chapter were evaluated the thermoplastic carbon fiber composites/PEI with respect to its performance, through two different processing technics: hot compression molding and aqueous suspension prepregging. The processing technic by hot compression molding can be considered mastered, in terms of obtaining composites, that is why, did not have a further explanation in this study, as it was the aqueous suspension prepregging, which is about the recent procedure and there is a few studies about this in the literature. Thus, much attention was given to the development, knowledge and description of this process.

Firstly, the carbon fiber fabric used in this study were superficially treated with nitric acid, based on the previews studies (Burakowski, 2001; Nohara, 2005), aiming the polar groups insertion on the fiber surface and pitting, responsible for a better mechanical anchorage of the matrix on the reinforcement. The treated carbon fibers were used only in the manufacture of composite type BTDA/ODA.

All the phases that comprise the method of aqueous suspension prepregging were analyzed; started by analyzing the different suspension behavior under pH. This investigation method had the objective to verify the suspensions that presented some behavior changes: color changing due to the pH or the deposition of the most polymeric matrix in the container, that could indicate the interaction or not of the PAAs to the polymeric matrix. Did not present changes in the analyzed systems, thus it was chose to use the pH 12. The best results were obtained for the composites PEI-PMDA/ODA, PEI-BTDA/ODA/fabric treated with HNO3 which presented a substantial increase in the ILSS values (18.1 and 21.9% respectively) compared to composite without interphase. This occurred probably, due to a interdiffusion between the PEI chains and the PIs, when found in oligomer form, in the polymeric suspension. The pre-impregnated PI-PMDA/ODA and PI-BTDA/ODA/treated fabrics presented the highest matrix adherence level to fiber fabrics, it was evidenced through a "polymeric film" on the fabric surface and a slight difficulty to be removed.

The 7 composites lots were analyzed to the fracture surface after the interlaminar shear strength. In general, it was observed shear failure occurred at the bulk the matrix and not in interface, indicating that this region was not responsible for the composites rupture. It was also a good wettability of fibers by the polymeric matrix has not been identified large quantities of dry fibers in the fracture surfaces of laminates.

5. Acknowledgment

The authors would like to thank FAPESP (São Paulo Research Foundation) (Process N. 00/15107-5), CNPq (National Council for Scientific and Technological Development) (Process N. 151929/2010-6 and 305478/2009-5) and CAPES (Coordination for the Improvement of Higher Level Personnel) (Process N. 0023/08-6) for the financial support.

6. References

Asao, K.; Saito, H. (2001). Polyamic acid and polyimide microfine particles and process for their production. United States Patent 6,187,899.

Bessonov, M. I. & Zubkov, V. A. (1993). Polyamic acids and polyimides, synthesis, transformations, and structure. CRC Press, ISBN 0849367042, Boca Raton.

Brink, A. E.; Lin, M. C. & Riffle, J. S. (1993). A high-performance electrostatic stabilizer for poly(ether ether ketone) particles. Chemistry of Materials. Vol. 5, No. 7, (July 1993), pp. 925-929, ISSN 0897-4756.

Brodowsky, H. M.; Jenschke, W. & Mäder, E. (2010). Characterization of interphase properties: microfatigue of single fibre model composites. Composites: Part A. Vol. 41, (July 2010), pp. 1579-1586, ISSN 1359-835X.

Burakowski L. Estudo da interface de compósitos termoplásticos estruturais processados a partir de fibras de carbono com superfícies modificadas. (2001). MSc Thesis. São José dos Campos, Brazil: Instituto Tecnológico de Aeronáutica.

Chuang, S. L.; Chu, N.-Jo; Whang, W. T. (1990). Effect of polyamic acids on interfacial shear strength in carbon fiber/aromatic thermoplastics. Journal of Applied Polymer Science. Vol. 41, No. 1-2, (1990), pp. 373-382, ISSN 1097-4628.

Chung, D. D. L. (2000). Fibrous composites Interfaces studied by electrical resistance measurement. Advanced Engineering Materials. Vol. 2, No. 12, (June, 2000), pp. 788-796, ISSN 1438-1656.

Cogswell, F. N. (1992). Thermoplastic Aromatic Polymer Composites. A study of the structure, processing and properties of carbon fiber reinforced polyetheretherketone and related materials. Butterworth-Heinemann Ltd., isbn 9780750610865, Great Britain.

Gao, S.-L. & Kim, J.-K. (2001). Cooling rate influences in carbon fibre/PEEK composites. Part II: interlaminar fracture toughness. Composites Part A. Vol. 32, No. 6, (June 2001), pp. 763-774, ISSN 0032-3861.

Gardner, S. H. An investigation of the structure-property relationships for high performance thermoplastic matrix, carbon fiber composites with a tailored polyimide interphase.

(1998). Doctoral Thesis. Blacksburg: Virginia Polytechnic Institute and State University.

Ghosh, M. & Mittal, K. L. (1996). Polyimides: fundamentals and Applications. Marcel Dekker, ISBN 9780824794668, New York.

Jenkins, M. J. (2000). Relaxation behaviour in blends of PEEK and PEI. Polymer. Vol. 41, No. 8, (August 2000), pp. 6803-6812, ISSN 0032-3861.

Kong, X. et al. (1996). Miscibility and crystallization behavior of poly(ether ether ketone)/polyimide blends. Polymer. Vol. 37, No. 9, (April 1996), pp. 1751-1755, ISSN 0032-3861.

Lee, J.-S. & Kang, T.-J. (1997). Changes in physico-chemical and morphological properties of carbon fiber by surface treatment. Carbon. Vol. 35, No. 2, (May 1998), pp. 209-216, ISSN 0008-6223.

Mittal, K. L. (1984). Polyimides: Synthesis, characterization, and applications. (Vol. 1), Plenum Press, ISBN 9780306416736, New York.

Nohara, L. B. et al. (2005). Evaluation of Carbon fiber surface treated by chemical and cold plasma processes. Materials Research, Vol. 8, No. 3, (September 2005), pp. 281-286, ISSN 1516-1439.

Nohara, L. B. (2005). Estudo da interface de compósitos termoplásticos estruturais processados a partir de moldagem por compressão a quente e suspensões poliméricas. (2005). Doctoral Thesis. São José dos Campos, Brazil: Instituto Tecnológico de Aeronáutica.

Nohara, L. B. et al. (2007). Otimização da interface/interfase de compósitos termoplásticos de fibra de carbono/PPS pelo uso do poli(ácido âmico) do tipo BTDA/DDS. Polímeros: Ciência e Tecnologia, Vol. 17, No. 3, pp. 180-187, ISSN 1678-5169.

Nohara, L. B. et al. (2010). Processing of high performance composites based on PEEK by aqueous suspension prepregging. Materials Research, Vol. 13, No. 2, (May 2010), pp. 245-252 , ISSN 1516-1439.

Offringa, A. R. (1996). Thermoplastic composites - rapid processing applications. Composites, Vol. 27 A, pp. 329-336, ISSN 0032-3861.

Pittman, C. U.; He, G. -R.; Wu, B. & Gardner, S. D. (1997). Chemical modification of carbon fiber surfaces by nitric acid oxidation followed by reaction with tetraethylenepentamine. Carbon, Vol. 35, No. 3, (January 1997), pp. 317-331, ISSN 0008-6223.

Pompe, G. Mäder, E. (2000). Experimental detection of a transcrystalline interphase in glass-fibre/polypropylene composites. Composites Science and Technology. Vol. 60, No. 11, (August 2000), pp. 2159-2167, ISSN 0266-3538.

Reifsnider, K. L. (1994). Modelling of the interphase in polymer-matrix composite material system. Composites. V. 25, No. 7, (March 1994), p.461-469, ISSN 1359-835X.

Saeed, M. B. & Zhan, M.-S. (2006). Effects of monomer structure and imidization degree on mechanical properties and viscoelastic behavior of thermoplastic polyimide films. European Polymer Journal. Vol. 42, No. 8, (August 2006), pp. 1844-1854, ISSN 0014-3057.

Srinivas, S. et al. Semicrystalline polyimides based on controlled molecular weight phthalimide end-capped 1-3-bis(4-aminophenoxy)benzene and 3,3',4,4'-biphenyltetracarboxylic dianhydride: synthesis, crystallization, melting, and thermal stability. Macromolecules. Vol. 30, No. 4, (February 1997), pp. 1012-1022, ISSN 0024-9297.

Sroog, C. E. Polyimides. Prog. Polym. Sci. Vol. 16, pp. 561-694, 1991.

Texier, A., et al. (1993). Fabrication of PEEK/carbon fibre composites by aqueous suspension prepregging. Polymer. Vol. 34, No. 4, (February 1993), pp. 896-906, ISSN 0032-3861.

Yu, T. H.; Davis, R. M. (1993). The effect of processing conditions on the properties of carbon fiber-LaRC TPI composites made by suspension prepregging. Journal of Thermoplastic Composite Materials. Vol. 6, No. 1, pp. 62-90.

Yue, Z. R.; Jiang, W.; Wang, L.; Gardner, S. D. & Pittman Jr., C. U. (1999). Surface characterization of electrochemically oxidized carbon fibers. Carbon, Vol. 37, No. 11, (October 1999), pp. 1785-1796, ISSN 0008-6223.

Zielke, U.; Hüttinger, K. J. & Hoffman, W. P. (1996). Surface-oxidized carbon fibers: I. Surface structure and chemistry. Carbon, Vol. 34, No. 8, (January 1996), pp. 983-998, ISSN 0008-6223.

6

Thermoplastic Nanocomposites and Their Processing Techniques

Sajjad Haider[1*], Yasin Khan[2], Waheed A. Almasry[1] and Adnan Haider[3]
[1]*Chemical Engineering Department,*
College of Engineering, King Saud University, Riyadh
[2]*Electrical Engineering Department, College of Engineering, King Saud University, Riyadh*
[3]*Department of Chemistry, Kohat University of Science and Technology, Kohat*
[1,2]*Saudi Arabia*
[3]*Pakistan*

1.Introduction

Nanotechnology is one of the most up-to-the-minute areas in essentially all technical disciplines of chemistry, electronics, high-density magnetic recording media, sensors and biotechnology,etc. During the last decade, due to the emergence of a new generation of high-technology materials, the number of research groups involved in nanotechnology has increased exponentially covering a broad range of topics such as microelectronics (now known as nanoelectronics, because the critical dimension scale for modern devices has decreased to and /or below 100 nanometer (nm) (Paul & Robeson 2008)), polymer-based biomaterials (Haider et al.,2007), nanoparticle drug delivery (Omer et al., 2011), miniemulsion particles (Zhang et al ., 2007; Wu et al., 2010)], layer-by-layer self-assembled polymer films (Lee & Cui 2009), electrospun nanofibers (Haider & Park, 2009; Haider et al., 2010), imprint lithography (Stephen et al., 1996), polymer blends and nanocomposites (Kim et al .,2005). The dimensional shift of materials from micro to nano produced theatrical changes in their physical properties and one such change is large surface area for a given volume (Haider et al.,2007). Therefore nanoscale materials can have substantially different properties from their corresponding large-dimensional materials of the same composition. Surface area per unit volume is inversely proportional to diameter, thus, the smaller the diameter, the greater the surface area per unit volume (Luo & Danie, 2003). Common particle geometries and their respective surface area-to-volume ratios are shown in Figure 1.

The surface area/volume ratio for the fiber and layered nanomaterials is subjugated by the first term (2/r or2/t) in the equation. The second term (2/l and 4/l) has a very small influence as compare to the first term (Frazana et al., 2006). Hence, understandably, altering the particle diameter, layer thickness, or fibrous material diameter from the micrometer (μm) and nm range, will affect the surface area/volume ratio by three orders of magnitude (Thostenson et al., 2005). The nanomaterials, which are under investigation most recently are, nanoparticles (gold (Au), silver (Ag), iron oxide (Fe_3O_4) titanium oxide (TiO_2), silicon oxide (SiO_2), and quantum dots

* Corresponding Author

(QDs),etc.), carbon nanotubes(CNTs) (SWCNTs (single walled carbon nanotubes), DWCNTs (double walled carbon nanotubes), MWCNTs (multi walled carbon nanotubes), fullerenes, electrospun nanofibers (Haider et al., 2011) and nanowires (Tsivion et al., 2011).

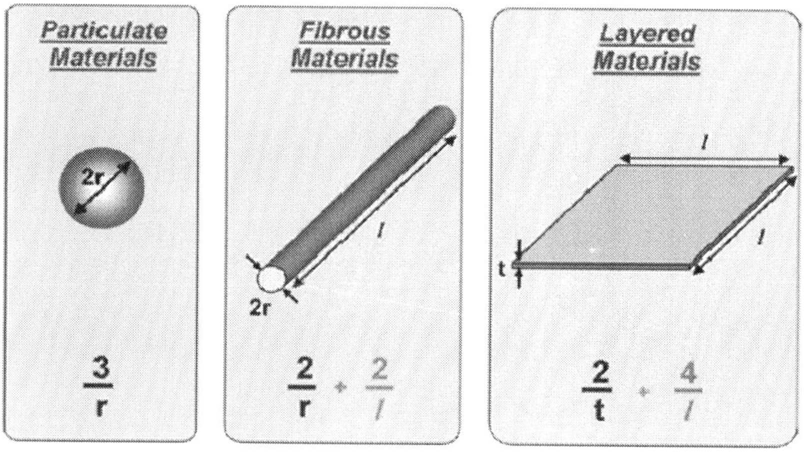

Fig. 1. Nanoparticles geometries and their surface area to volume ratio (Thostenson et al., 2005)

These materials are generally classified by their geometries into three broad classes, and these are particle (gold (Au), silver (Ag), iron oxide (Fe_3O_4) titanium oxide (TiO_2), silicon oxide (SiO_2), etc), layered (materials with nm thickness, high aspect ratio (30–1000) and plate-like structure ,e.g organosilicate), and fibrous (CNTs and electrospun nanofibers, etc) materials(Schmidt et al., 2002; Alexandre & Dubois2000). Nanomaterials provide reinforcing efficiency due to their high aspect ratios, when mixed with other matrix to produce composite material (Luo and Danie., 2003). Composite with different properties could be obtained and this mainly depends on the components (particles, layered or nanofibrous material, cation exchange capacity, and polymer matrix) and the method of preparation [Park et al., 2001). The interest in polymer matrix based nanocomposites has emerged initially with interesting observations involving exfoliated clay and more recent studies with CNTs, carbon nanofibers, exfoliated graphite (graphene), nanocrystalline metals and a host of additional nanoscale inorganic filler or fiber modifications. Nanotechnology is not new to polymer science as prior studies involving nanoscale dimensions before the age of nanotechnology were not particularly referred to nanotechnology until recently (Paul & Robeson 2008). In the last two decade, there has been increasing interest in the development of polymeric composites, where at least one of the phases is dimensionally in the nm size range (Mirkin, 2005). The story of nanotechnology and nanocomposites is manifested by several significant events. In 1980s the heave in nanocomposites development, has been facilitated to a great extent by the invention/introduction of powerful characterization tools i.e., scanning tunneling microscopy (STM), atomic force microscopy (AFM) and scanning probe microscopy (SPM). These tools made the researchers capable to see the nature of the surface structure and manipulate individual atoms and molecules on solid surfaces. (Royal

Society & the Royal Academy of Engineering, 2004). At the same time, the speedy development of computer technology has made it easier to characterize and predict properties at the nanoscale *via* modeling and simulation (Mirkin, 2005). In general, the unique nanomaterial's characteristics (such as size, mechanical properties, and low concentrations) coupled with the advanced characterization and simulation techniques, have generated much curiosity in researchers for studying nanocomposites. Furthermore many polymer nanocomposites were fabricated and processed in ways, which were similar to those of conventional polymer composites, making them predominant from manufacturing point of view. Nature has mastered in the fabrication of nanocomposites by utilizing natural reagents and polymers (such as carbohydrates, lipids, and proteins) and as always the case researchers have learned from nature. Nature makes strong nanocomposites such as bones, shells and wood,etc by mixing two or more phases such as particles, layers or fibers, where at least one of the phases is in the nanometer size range. The quotation "Nature is a master chemist with incredible talent' by Oriakhi explains all (Paul & Robeson 2008). In recent years, preparation of hybrid organic-inorganic composites has attracted much attention since (such hybrids may show controllable properties such as optical, electrical and mechanical behaviors) by combining the properties of both organic and inorganic compounds (Chiang et al., 2003), e.g. Nylon-6 nanocomposite have shown pronounced improvement of thermal and mechanical properties with very small amount of nanofiller loading resulted in a pronounced improvement of thermal and mechanical properties (Kojima et al., 1993; Saeed et al., 2009). This is not always the case, as the properties of nanocomposite materials depend not only on the properties of the individual parent materials but also on the morphology and interfacial characteristics. Due to the huge potential of the nanotechnology, federal funding for nanotechnology research and development in United State of America (USA) has considerably increased from $464 million since the inception of National Nanotechnology Initiative (NNI) in 2001 to $982 million in 2005 (Paul & Robeson 2008) and, this amount is expected to increase to $2.1 billion in 2012 (http://www.nano.gov/about-nni/what/funding). These incredible funding in nanotechnology are focused more on efficient manufacturing processes and the production of novel nanomaterial based products for a wide range of applications.

2. Thermoplastic polymers and their nanocomposites

Plastics are the basic ingredients of animal and plant life and form a part of the larger family called polymers. They offer advantages such as low density, lightness, transparency, resistance to corrosion and colour,their applications vary from domestic articles to medical instruments (Lia et al., 2000). Plastic are broadly classified into; (i) thermoplastics and (ii) thermosetting plastics. Both are long chain-like molecules but differs in their bonding. In thermoplastics the long chain molecules are held together by weak vander waal forces where as in thermosetting, the long chain molecules are held together by strong bonds. Some of the common thermoplastic polymers are; acrylonitrile butadiene styrene (ABS) [Zheng et al., 2004), Acrylic (PMMA) (Khaled et al ., 2007), Ethylene vinyl alcohol (EVOH) (Artz et al., (2002), Polyacrylonitrile (PAN)(Zhang et al., 2009), Polyamide-imide (PAI) (Ma et al., 2010),Polyamide (PA or Nylon)(Bourbigot et al., 2002), Polyethersulfone (PES) (Wang et al 2011), Polypropylene (PP)(Lee et al., 2009) Polystyrene (PS) (Meincke et al.,2003) Polytrimethylene terephthalate (PTT) (Hu et al., 2004) and Polyvinyl chloride (PVC) (Awad et al., 2009), etc. As mentioned above the weak vander waal force can be overcome at high temperatures, resulting in a homogenized viscous liquid, which can be further moulded into various shapes. After cooling, thermoplastic polymers retained their newly reformed shape. Some of the common applications of thermoplastics are; bottles, cable insulators, tapes, blender and mixer bowls, medical syringes, mugs, textiles, packaging, and parts for common household appliances. Nanomaterials reinforced thermoplastic nanocomposites have shown significant growth, as research on the development of novel reinforcing methods intensified. Several methods such as melt blending, solution mixing, coating (Kalaitzidou et al., 2010), in-situ polymerization (Alexandre &Dubois, 2000), nanoinfusion, etc (Lentz et al 2010) are reported in the literature mostly aiming on the dispersability of the nanomaterials. The nanocomposites prepared using the above mentioned methods has a huge potential for applications in automotive (seat frames, battery trays, bumper beams, load floors, front ends, valve covers, rocker panels and under engine covers, etc) (Garces et al., 2000), aerospace (missile and aircraft stabiliser fins, wing ribs and panels, fuselage wall linings and overhead storage compartments, ducting, fasteners, engine housings and helicopter fairings,etc) (Zhao et al., 2010), optical devices (Ritzhaupt-Kleissl et al., 2006), electrical and actuator devices (Koerner et al., 2004), and as flame retardant. The fabrication of thermoplastic nanocomposites with CNTs, Caly and graphene are discussed in detail as below sections.

2.1 Clay (silicates)-based nanocomposites

After the ground braking discoveries by Toyota research group (preparation of Nylon-6 (N6)/montmorillonite (MMT) nanocomposite) (Paul & Robeson 2008) and Vaia et al. (dry melt-mixing of polymers with layered silicates) (Vaia et al., 1994), research on polymer/clay nanocomposites accelerated, both in industry and in academic worlds. Since then a number of groups studied the preparation of thermoplastics/silicates nanocomposites e.g. ethylene-vinyl alcohol copolymer/clay (melt blending, solution-precipitation (Artzi et al., 2002; Jeong et al., 2005)), PAN/silicate (in situ polymerization (Choi et al., 2004)), PP/clay (melt-mixing and melt compounding (Parka et al., 2008)) and PS/ clay (in situ polymerization, melt intercalation and solution casting (Panwar et al., 2011)),etc., and found significant improvements in their material properties. Based on the interaction of silicates with polymer, polymer/layered silicate nanocomposites are classified into(Figure 2); (a) intercalated nanocomposites (inserted polymer chains in the layered silicate structure in a crystallographically regular fashion, irrespective of the polymer to clay ratio with a repeat distance of few nm), (b) flocculated nanocomposites, (intercalated and stacked silicate layers flocculated to some extent due to the hydroxylated edge-edge interactions of the silicate layers), and (c) exfoliated nanocomposites (separated individual silicate layers in the polymer matrix; the distance depends on the clay loading) (Ray & Okamoto, 2003) . The strong interactions of polymer with layered silicate in polymer/silicates nanocomposites lead to the dispersion of organic and inorganic phases at the nm level, resulting in improved tensile properties (Haider and Park 2009), high moduli (Vaia et al., 1999), increased strength and heat resistance, (Pattanayak et al., 2005), thermal stability (Xu et al., 2004),decreased gas permeability (Kojima et al., 1993), and flammability (Gilman et al., 2000), and increased biodegradability (Gloaguen et al., 2001), not shown by their conventionally filled micro-counterparts (Ray & Okamoto 2003).

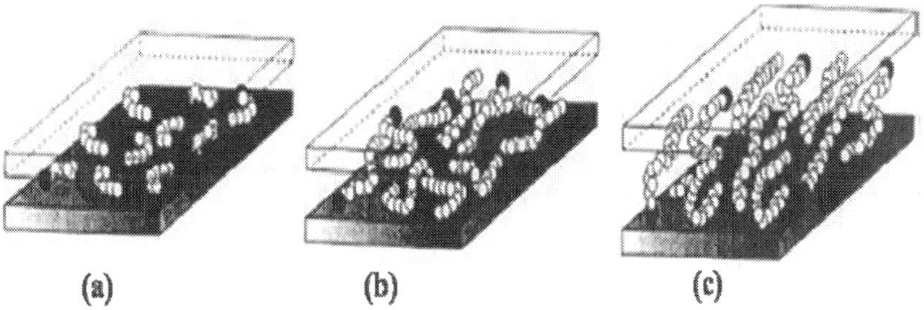

Fig. 2. Interaction of silicates with polymer (a) intercalation, (b) flocculation and (c) exfoliation (Vaia et al., 1994).

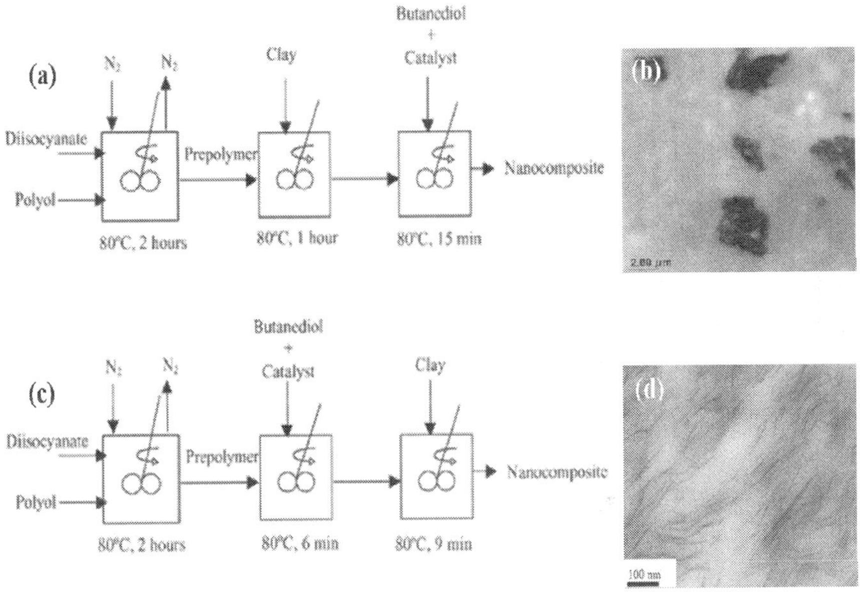

Fig. 3 Schematics of clay/polyurethane nanocomposite preparation and their respective TEM images; (a) method I and (b) TEM image of 4 wt% clay/polyurethane nanocomposite prepared by method I, (c) method II and (d) TEM image of 5 wt% clay/polyurethane nanocomposites prepared by method II.

Pattanayak and Jana prepared thermoplastic polyurethane and reactive silicate clays nanocomposites by bulk polymerization. They used two different methods as shown in Figure 3. In the first method pre-polymer and in the second method chain-extended polymer molecules with residual -NCO groups were involved in tethering reactions with clay during mixing. The morphologies studies of the nanocomposites showed that both shear stress of mixing and the reaction between polymer and clay are the main forces that determine the exfoliation of clay. In first method (Figure 3a) more clay-tethered to polymer chains, but the clay particles did not exfoliate (Figure 3b) due to low shear stress of mixing. In the second method (Figure 3c) due to the high shear stress of mixing and good reaction, the clay particles are well exfoliated to the scale of individual clay layers (Figure 3d). In the later case best improvement in tensile properties was observed (Pattanayak & Jana, 2005). Gloaguen et al., studied the preparation of nylon 6/clay hybrids *via* in situ polymerization, to obtain a nylon matrix strongly bonded to the delaminated clay platelets and polypropylene/organophilic clay *via* melt dispersion of organophilic clay in polypropylene, to obtain reduced degree of polymer-clay interaction. Extensive cavitational behavior was observed from plasticity results, while retaining a fairly large strain at break, as long as deformation is performed above the glass transition temperature of the matrix. In case of nylon6, it was clear that the usual shear banding plastic deformation mode was altered in its initiation step. Localised interfacial damage promoted extensive polymer matrix brillation and fracture occurred predominantly in areas where delamination of the clay platelets was not fully achieved (Gloaguen & Lefebvre, 2001).

Heterocoagulation method was also employed to prepare PMMA/clay nanocomposites. In the first step a cationic PMMA emulsion was prepared by emulsion polymerization using a cationic initiator in the presence of free surfactant, followed by mixing with aqueous clay slurries of montmorillonite (MMT), synthetic hectorites and fluorohectorites, in the second step. Based on hetero and homocoagulation processes three types of morphologies (Figure 4 (i. exfoliated (a, b, and c), ii. mixed (d) and iii. intercalated (e))_ were predicated. TEM images confirmed the predicted morphologies (Figure 4a, b,d,e). The enhanced thermal stability for these systems was attributed partially to barrier properties (Gloaguen & Lefebvre, 2001).

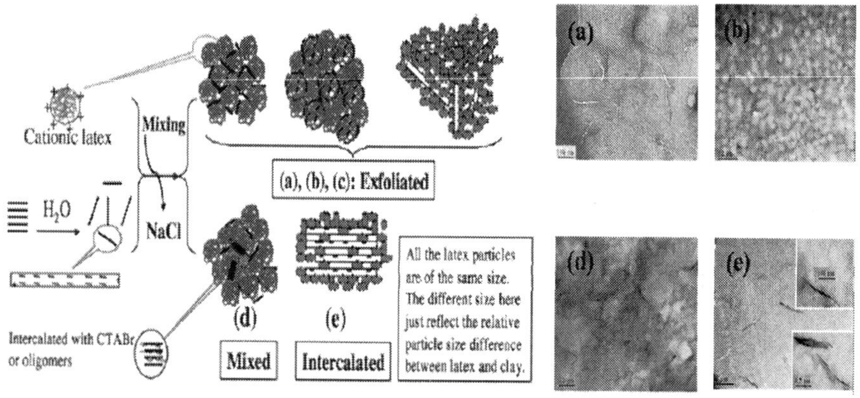

Fig. 4. Schematic representation of the predicted morphologies for coagulation process and TEM images confirming the predicted morphologies (Vaia et al., 1999).

2.2 CNTs -based nanocomposites

Synthetic polymers lack the necessary stiffness (with Young's modulus, 3000- 4000 MPa for amorphous and 3000 MPa for semi-crystalline polymers) for many engineering applications. Polymer industry is therefore in continuous search for new materials, which may not only potentially enhance the stiffness of synthetic polymer but could add some new properties (such as thermal stability, electric conductivity and air stability, etc.) at lower costs. CNTs with superb mechanical properties (Young's modulus 1TPa and tensile strength 100-150 GPa), electrical conductivity as high as 10^8 S/m, high aspect ratio, small diameter, light weight, air stabilities and thermal stabilities could be one such material. Hence, CNTs have been extensively studied since the discovery by Iijima in 1991 (Ando, 2010). Having said this, pristine CNTs did not yet met early expectations; however their incorporation into polymer materials to prepare nanocomposites, have extend their applications, despite inconsistencies in properties due to (i) CNTs synthesis method and the resultant aspect ratio, (ii) their post-processing (such as purification), and (iii) the characteristics of polymer matrix and the nanocomposite fabrication procedure (such as solution blending, melt blending, or in situ polymerization) (Silva et al., 2011). It is therefore necessary to entertain a compromise between the properties and processing, since only low loadings of nano-filler are needed to reinforce the matrix, which might exhibit improved stiffness without scarifying toughness (Chang et al., 2006).

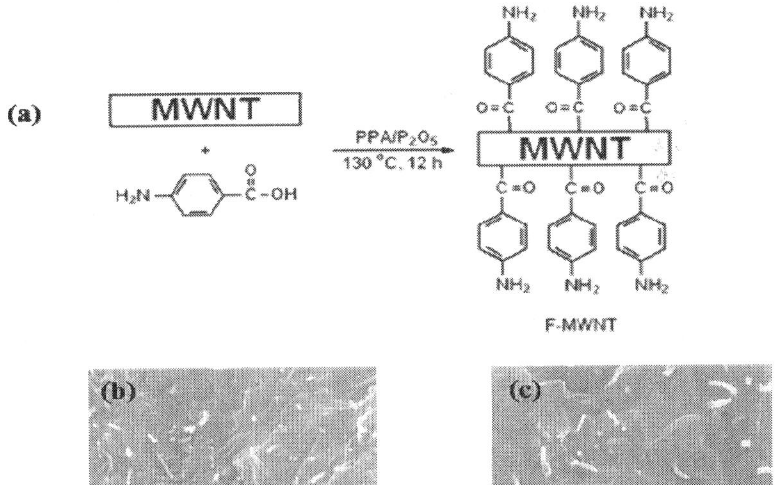

Fig. 5. (a) Schematic of the Side-wall functionalization of MWNTs (F-MWNTs) by Friedel-Crafts acylation, (b) SEM images of the fractured surfaces of a P-MWNT (5 wt%)/nylon-6 and (c) F-MWNT (5 wt%)/nylon-6 (Saeed et al., (2009)

CNTs reinforced thermoplastic nanocomposites have shown considerable growth with increase in the research on the development of novel reinforcing methods. Several methods have been reported in the literature to incorporated CNTs into the thermoplastic polymer matrix. These methods include, solution blending (PAN/MWCNT (Saeed &Park 2010), melt blending a (PMMA/MWCNT, (Yu et al., 2008), PP/SWCNT(Bhattacharyya et al., 2003)), in situ polymerization (Nylon/MWCNT (Saeed et al., 2009), PP/MWCNT (Koval'chuk et al., 2008), PMMA/MWCNT(Cui et al., 2009), PS/MWCNT (Lahelin et al., 2011)), solution casting (PAI/SWCNT(Kim et al., 2007) PTT/MWCNT (Xu et al., 2008)), and Solution

Polymerization (PS/MWCNT (Kim et al., 2007). The properties of the final matrix obtained by the above mentioned methods, depends largely on the dispersion of the CNTs in the polymer matrix because nanoscale diameter and the resultant high aspect ratio gives CNTs very larger surface area, which induces significant entanglement between nanotubes. This entanglement of CNT must be worked out to realize its high potential in practical applications. Various approaches have been use to obtain good dispersion (e.g. chemical functionalization and physical wrapping (Saeed et al., (2009), etc) of the CNTs in polymer matrix. Khalid et al prepared Nylon/MWCNTs nanocomposite by in situ polymerization with both pristine (Figure 5b) and chemically functionalized MWCNTs (Figure 5c). The MWCNTs were functionalized by Friedel-Crafts acylation (Figure 5a), which introduced aromatic amine (COC_6H_4-NH_2) groups onto the side wall. Homogenous dispersion of the MWCNTs (Figure 5c) was observed in nylon, which results in an increase in both thermal and mechanical properties for the nanocomposite as compared to the neat polymer (Saeed et al., 2009).

2.3 Graphite-based nanocomposites

Graphene, a monolayer of sp^2 hybridized carbon atoms arranged in a two-dimensional lattice, is a cheap and multifunctional material (Potts et al., 2011). Graphene is considered to be better nanofiller compared to CNTs and other conventional nanofillers, as it improve the mechanical, thermal, and electrical properties of the nanocomposites to a great extent with a very small loading. However, the dispersion of pristine graphene in polymer matrices, as is the case with nanoscale materials, is very poor (Kuilaa et al., 2011). Poor dispersion drastically effect mechanical, thermal, and electrical properties of the nanomaterial reinforced nanocomposites. Currently work is intensified on making these nanofiller homogeneously dispersible in the polymer matrix. Different types of nano graphite forms, such as expanded graphite (forcing the crystal lattice planes apart due to the insertion of some foreign particles), intercalated (insertion of metal between lattice planes) and surface modified graphene have been used to obtain homogeneous dispersion and produce nanocomposites with improved physicochemical properties. The mechanism of polymer-graphene interactions in polymer/graphene nanocomposites is mainly governed by (i) polarity, (ii) molecular weight, (iii) hydrophobicity, (iv) polymer functionalities, (v) graphene functionalities and (vi) graphene-solvent interaction. In the literature three main approaches are devised for incorporating polymer between graphene layers; (i) In situ intercalative polymerization, (ii) solution intercalation and (iii) melt intercalation. The first methods, is commonly used for the preparation of the homogeneously dispersed graphene. In this method pristine or expended, intercalated or surface modified graphene is first swollen within the liquid monomer and after the diffusion of a suitable initiator, polymerization is initiated either by heat or radiation (Kuilla et al., 2010). A number of thermoplastic polymers/gaphene nanocomposites with much improved mechanical, thermal and electric properties were prepared by this method, e.g PMMA/graphene (enhanced storage moduli , glass transition temperatures and thermal stability (Kuilla et al., 2011)), PP/graphene (effective dispersion and high electic conductivity (Huang et al., 2010)), Nylon/graphene (increased Young's modulus (Xu & Gao 2010)) and PS/graphene (improved thermal properties (Patole et al., 2010)), etc. Solution intercalation, not so common, is based on the adsorption of polymer or pre-polymer onto graphene layers. During the process polymer or pre-polymer is solubilized in solvent system, followed by graphene swelling of pristine or modified graphene layers in the same solution. When the polymer or pre-polymer adsorbs onto the delaminated sheets, the solvent is evaporated and a sandwiched nanocomposites is formed (Kuilla et al., 2010). As mentioned, this method is rarely used for the preparation of thermoplastic polymer/graphene nanocomposites, only PP/graphene nanocomposite is reported recently (increased electric conductivity (Kalaitzidou et al., 2007)). In the melt intercalation technique, no solvent is required and graphite or graphene or modified graphene is mixed with the polymer matrix in the molten state (Kuilla et al., 2010), e.g. PTT/graphene (substantially improved thermal stability and dynamic mechanical moduli (Li &Jeong 2011)).

2.4 Processing techniques

Processing conditions are the factors, which need to be optimized to enhance the performance of thermoplastic polymers and / or their nanocomposites. The processing of polymeric materials (such as plastics, elastomers and composites, etc) is characterized by a wide variety of distinct techniques such as extrusion (Vaia et al., 1994), film blowing (Golebiewski et al., 2008), sheet thermoforming (Feng et al., 2009), blow molding (http://www.petmachine.in/type_of_blow_moulding.htm), and injection molding (Chandra et al., 2007), etc. Extrusion is the continuously shaping of a fluid polymer and/or its nanocomposite through the orifice of a suitable tool (die) (Figure 6a), followed by solidifying it into extrudate of constant cross section. The feed material is usually thermoplastics powder or pellets. In this process, the feed material is heated first to a fluid state *via* a screw extruder, followed by pumping into the die, which is finally solidified by cooling. Extrusion products are often subdivided into groups that include filaments of circular cross-section, profiles of irregular cross section, axis-symmetric tubes and pipes, and flat products such as films or sheets. Another technique used for the processing of thermoplastics into tubular product several times of its initial diameter that can be use directly or made into film, is known as film blowing (Figure 6b). Films produced by the film blowing process are widely used for agricultural, construction, and industrial applications, including covers for silage, greenhouses, chemical/solar ponds, flat cars, etc., or for a variety of packing applications, such as wrapping, can lining, fabricated bags such as garbage, etc (http://www.whatisplastic.com/?cat=9).

Blow molding (Figure 7) is the simplest type of molding, in this process, a hot tube of plastic material is dropped from an extruder and captured in a water cooled mold halves. Once the molds are closed, air is injected through the top or the neck of the container; just as if one were blowing up a balloon. The material solidifies into a hollow product. Packing is the major area of application of small to medium-size disposable blow molded products. Blow molded containers are also used for cosmetics, toiletries, pharmaceutical, medical packaging and a variety of household products.

The injection molding process involves the rapid pressure filling of a specific mold cavity with a molten material, followed by the solidification of the material into a product. This process is used for molding thermoplastics, thermosetting resins, and rubbers, etc. Injection molding of thermoplastics can be classified into a several stages. At the plasticity stage, the feed unit operates as an extruder, melting and homogenizing the material in the screw/barrel system. The screw, however, is allowed to pull in, to make a reservoir for the molten materials. At the injection state, the screw works as a ram, which transfer the molten material rapidly from the reservoir to the cavity between the two halves of the closed mold.

Since the mold is kept at a temperature below the solidification temperature of the material, it is essential to inject the molten material rapidly, to ensure filling of the cavity completely. The cavity pressure rises rapidly during the filling stage, which is followed by the holding stage. A high holding/packing pressure is normally exerted, to partially compensate for the thermal contraction of the material upon cooling. After the cooling stage, the mold can be opened and the solid product removed (http://www.petmachine.in/type_of_blow_moulding.htm).

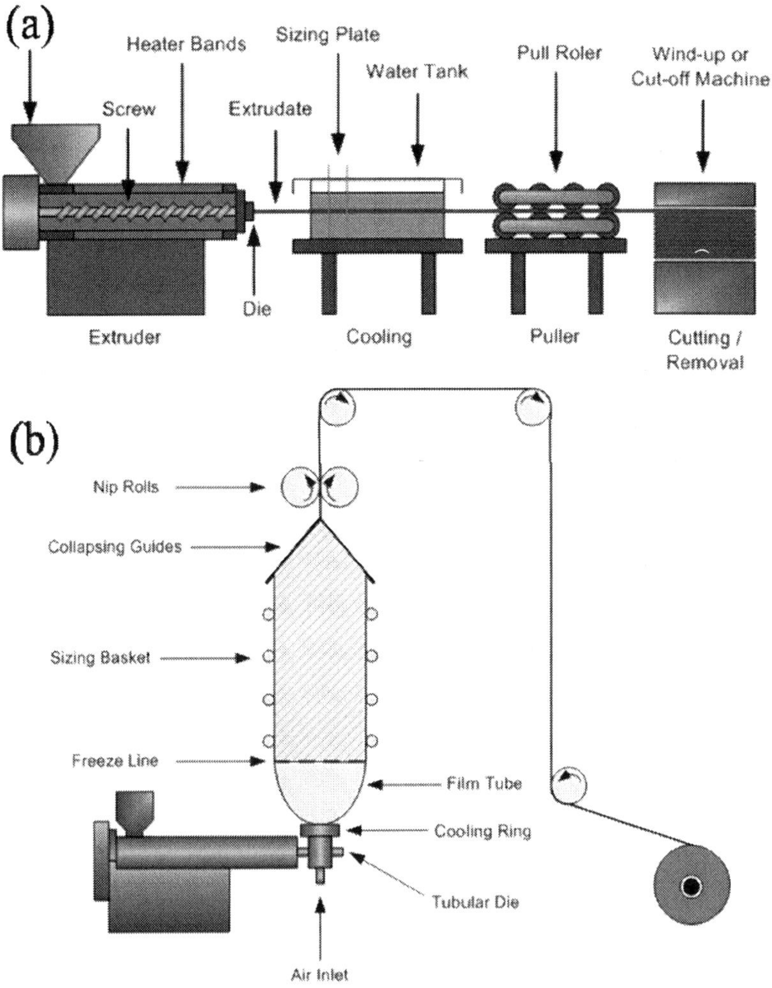

Fig. 6. Schematic for processing of thermoplastic polymers and its nanocomposite (a) Extrusion and (b) film blowing

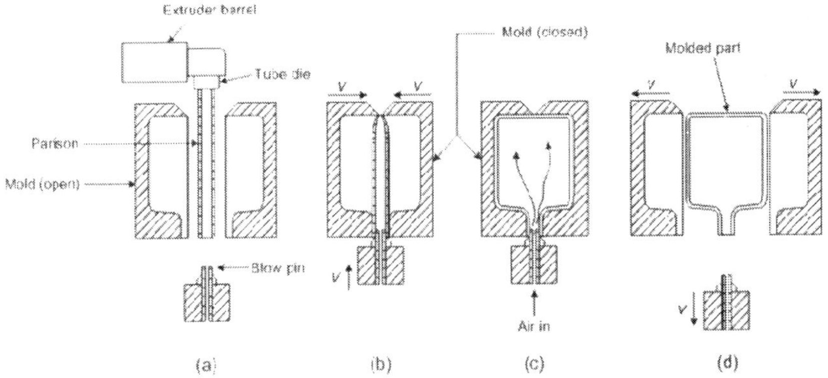

Fig. 7. Extrusion blow molding: (a) extrusion of parison; (b) parison is pinches at the top and sealed at the bottom around a metal blow pin as the two halves of the mold come together; (c) the tube is inflated so that it takes the shape of the mold cavity; and (d) mold is opened to remove the solidified part (http://www.petmachine.in/type_of_blow_moulding.htm).

At high pressures, a polymer melt is compressible; allowing additional material to be packed in the mold cavity after mold filling is complete. This is necessary to reduce non-uniform part shrinkage, which leads to part warpage. Excessive packing results in a highly stressed part and may cause ejection problems whereas insufficient packing causes poor surface, sink marks, welds and non-uniform shrinkage. All thermoplastics are, in principle, suitable for injection molding, but since fast flow rates are needed, good fluidity thermoplastic are normally preferable. The pressure distribution inside the mold cavity changes with distance from the inlet gate. Further away from the gate, pressure rises slowly and it decays quicker than at the points closer to the gate. The pressure in the mold cavity should be more uniform to minimize part warpage. A major disadvantage of injection molded products is the incorporation of fine details such as bosses, locating pins, mounting holes, ribs, flanges, etc., which normally eliminates assembly and finishing operations. (Schey, 1987; Strong, 2000). Thermoforming is another very useful process for producing various devices. Thermoforming involves the heating of thermoplastic sheet above the glass transition temperature T_g for non-crystallizing thermoplastics or near the melting temperature T_m for crystallizing polymers, followed by forming of the softened material into a desired shape by cooling. Products made by sheet thermoforming include skin and blister packs, individual containers (for jelly or cream, vials), cups, tubs, trays and lids. Larger products are generally made from cut sheets at much slower rates; the heating stage often is the limiting factor. Transparent products, such as contoured windows, skylights and cockpit canopies, are often made by this method (http://www.thermopro.com/FormingMethods.html).

3. Characterization techniques for nanocomposites

Characterization is critical for analyzing the physical and chemical properties of polymer nanocomposites. Various characterization techniques such as scanning electron microscope (SEM), transmission electron microscopy (TEM), Atomic force microscope (AFM), Raman and Fourier transform infra red (FT-IR) spectroscopy wide-angle X-ray diffraction (WAXD), small-angle X-ray scattering (SAXS), thermogravimetirc gravity analysis (TGA) and differential scanning calorimetry (DSC), have been reported in the literature. SEM is used pre-dominantly for imaging the surface texture of materials, however; it can also be used for particle sizing and aggregation. Conventional SEM requires UHV environment and dehydrated samples (dry samples), which lead to uncertainty in getting accurate data in case of wet samples. The problem has been overcome and nowadays the liquid (aqueous) surface and substances close to the surface can be imaged in in-situ material characterization while maintaining UHV conditions around the electron gun with commercially available emission scanning electron microscopy (ESEM). TEM is a entrenched direct electron imaging technique under ultrahigh vacuum (UHV) conditions for studying shape, morphology, particle size distributions and aggregation of nonmetallic and in particular of metallic nanomaterials, though in the former case emission field transmission electron microscope (EFTEM) might be used due to the energy filtering discrepancy (Haider et al ., 2011). AFM uses a sharp tip to scan across the sample for measuring three dimensions morphology of the surface. FTIR and Raman spectroscopy has also proved a useful probe of the functionalities in the polymer matrix and their interaction with the nanomaterials (Gajendran& Saraswathi 2008). WAXD is commonly used to probe the nanocomposite structure (Monticelli et al., 2007) and kinetics of the polymer melt intercalation (Li et al., 2003). In layered silicate nanocomposite, a fully exfoliated system is characterized by the absence of intensity peaks in WAXD pattern, which corresponds to a d-spacing of at least 6 nm (Lia et al., 2000). Therefore, a WAXD pattern concerning the mechanism of nanocomposite formation and their structure are tentative issues for making any conclusion. Small-angle X-ray scattering (SAXS) is typically used to observe structures on the order of 10A °or larger, in the range of 0.5^{o}–5° . The TEM, AFM, and SEM, are also required to characterize nanoparticle, carbon nanofiber dispersion, or distribution (Hussain et al., 2006). TGA is useful tool for analyzing the amount of the nanomaterials loaded in the matrix and the thermal stability of the nanocomposites whereas DSC provide us the information of the glass transition and melting temperature of the nanocomposites, the results can be compared with the neat matrix and conclusion could be drawn about the interaction of the nanomaterials and matrix.

4. Future outlook

The huge potential of nanocomposites in various disciplines of research and industrial applications is attracting increasing investment in many parts of the world from governments and business companies. The federal funding for nanotechnology research and development in United State of America (USA) has considerably increased from $464 million, since the inception of National Nanotechnology Initiative (NNI) in 2001, to $982 million in 2005 (Paul & Robeson 2008) and this amount is expected to increase to $2.1 billion in 2012 (http://www.nano.gov/about-nni/what/funding). Over 300 nanomaterials products were available in the market until 2007. Even if these numbers are below expectation, they highlight the tremendous technological and economical potential associated with polymeric nanocomposites including not only clay but also other inorganic and organic nanofillers, such as carbon nanotubes, SiO_2, SiC, and Si_3N_4. To prepare these nanomaterials on macroscale, various issues surrounding the incorporation of nanomaterials into polymer matrix, strategies for property improvement, and the mechanisms responsible for those property improvements still remain critical. Thus, polymeric nanocomposites expected to spread through all aspects of life in the mid and longer-term, similar to the way plastics did in the last century. Clearly a diverse range of sectors such as aerospace, automotive, packaging (particularly food but also solar cells), electrical and electronic goods, and household goods etc., will significantly profit from a new range of materials, offered by nanotechnology. In the short term (<5 years), the commercial impact may include inkjet markets, cosmetics, automotive (body moldings, engine covers), catalytic converts and computer chips, (Paul & Robeson 2008; Hussain et al., 2006), while in the medium term (>5- <15 years) (10 years), memory devices, biosensors for diagnostics, advances in lighting applications. The time-scale for automotive, aerospace, bio-nanotechnology is a long-term prospect (>15 years) as these are risk-averse sectors and thereby for large-scale production it is necessary to carry out strict testing and validation procedures (Hussain et al., 2006).

5. References

Ando Y.(2010). Carbon nanotube: Story in ando laboratory. Journal of Siberian Federal University Mathematics & Physics, 2010, 3(1), 3-22.

Awad W. H., Beyer G., Benderly D., Ijdo W. L., Songtipya P., Gasco M. M. J., Manias E., Wilkie C. A. (2009). Material properties of nanoclay PVC composites. Polymer, 50 (8) 1857-1867.

Artz N., Nir Y., Narkis M., Siegmann A. (2002). Melt blending of ethylene-vinyl alcohol copolymer/clay nanocomposites: Effect of the clay type and processing conditions. Journal of Polymer Science Part B: Polymer Physics, 40 (16) 1741-1753.

Alexandre, M., Dubois P. (2000). Polymer-layered silicate nanocomposites: preparation, properties and uses of a new class of materials. Materials Science and Engineering, 28 (1-2) 1-63.

Bhattacharyya A. R., Sreekumar T. V., Liu T, Kumar S., Ericson L. M., Hauge R. H., Smalley R. E. (2003). Crystallization and orientation studies in polypropylene/single wall carbon nanotube composite. Polymer, 44 (8) 2373–2377.

Bourbigot S., Devaux E., Flambard X. (2002). Flammability of polyamide-6/clay hybrid nanocomposite textiles. Polymer Degradation and Stability, 75 (2) 397-402.

Chiang C.L., Ma C.C. M., Wang F. Y, Kuan H. C. (2003). Thermo-oxidative degradation of novel epoxy containing silicon and phosphorous nanocomposites, European Polymer Journal, 39 (4) 825–830.

Cui L., Tarte N. H., Woo S. I. (2009). Synthesis and Characterization of PMMA/MWNT nanocomposites prepared by in situ polymerization with Ni(acac)2 catalyst. Macromolecules, 42 (22) 8649–8654.

Chandra A., Kramschuster A. J., Hu X., Turng L.-S. (2007). Effect of injection molding parameters on the electrical conductivity of polycarbonate/carbon nanotube nanocomposites. ANTEC, 2184-21.

Chang T.-E., Kisliuk A., Rhodes S. M., Brittain W. J., Sokolov A. P. (2006). Conductivity and mechanical properties of well-dispersed single-wall carbon nanotube/polystyrene composite. Polymer, 47 (22) 7740-7746.

Choi Y. S., Chung I. J. (2004). An explanation of silicate exfoliation in polyacrylonitrile/silicate nanocomposites prepared by in situ polymerization using an initiator adsorbed on silicate. Polymer, 45 (11) 3827-3834.

Feng L., Zhou Z., Dufresne A., Huang J., Wei M., An L. (2009). Structure and properties of new thermoforming bionanocomposites based on chitin whisker-graft-polycaprolactone. Journal of Applied Polymer Science, 112 (5) 2830–2837.

Frazan H., Medhi H., Masami O., Russell E.G. (2006). Review article: Polymer matrix nanocomposite processing. Manufacturing and application: An over view. Journal of Composite Material, 40 (7) 1511-1575.

Gajendran P., Saraswathi R. (2008). Polyaniline-carbon nanotube composites. Pure and Applied Chemistry, 80 (11) 2377–2395.

Golebiewski J., Rozanski A., Dzwonkowski J, Galeski A. (2008). Low density polyethylene-montmorillonite nanocomposites for film blowing. European Polymer Journal, 44 (2) 270–286.

Gloaguen J. M., Lefebvre J. M. (2001). Plastic deformation behaviour of thermoplastic/clay nanocomposites. Polymer, 42 (13) 5841-5847.

Garces J. M., Moll D. J., Bicerano J., Fibiger R., McLeod D. G. (2000).Polymeric Nanocomposites for automotive applications. Advance Material, 12(23) 1835-1839.

Gilman J. W., Jackson C. L., Morgan A. B., Harris J. R., Manias E., Giannelis E.P., Wuthenow M., Hilton D., Phillips S.H. (2000). Flammability properties of polymer-layered silicate nanocomposites. Propylene and polystyrene nanocomposites. Chemistry of Material, 12 (7) 1866–1873.

Haider S., Bukhari N., Haider A. (2011). *Risk associated with the use of nanomaterials*, Book Chapter, Bentham Science Publishers Ltd, USA (In-press).

Haider S., Al-Masry W A., Bukhari N., Javid M.(2010). Preparation of the chitosan containing nanofibers by electrospinning chitosan-gelatin complex. Polymer Engineering & Science, 50 (9) 1887–1893.

Huang Y., Qin Y., Zhou Y., Niu H., Yu Z.-Z., Dong J.-Y. (2010). Polypropylene/Graphene oxide nanocomposites prepared by in situ ziegler−natta polymerization. Chemistry of Materials, 22 (13) 4096–4102.

Haider S., Park S.Y.(2009). Preparation of the electrospun chitosan nanofibers and their applications to the adsorption of Cu(II) and Pb(II) ions from an aqueous solution. Journal of Membrane Science, 328 (1-2) 90-96.

Haider S., Park S.Y., Saee K., Farmer B. L. (2007). Swelling and electroresponsive characteristics of gelatin immobilized onto multi-walled carbon nanotubes. Sensors and Actuators B: Chemical, 124 (2) 517-528.

Hussain F., Hojjati M., Okamoto M., Gorga R. E. (2006). Review article: Polymer-matrix nanocomposites, processing, manufacturing. and Application: An overview, Journal of Composite Materials, 40 (17) 1511-1575.

Hu X., Lesser A. J. (2004). Non-Isothermal Crystallization of Poly(trimethylene terephthalate)(PTT)/Clay Nanocomposites. Macromolecular Chemistry and Physics, 205(5) 574–580.

http://www.petmachine.in/type_of_blow_moulding.htm

http://www.whatisplastic.com/?cat=9

http://www.thermopro.com/FormingMethods.html

http://www.nano.gov/about-nni/what/funding

Jeong H. M., Kim B. C., Kim E.H.(2005). Structure and properties of EVOH/organoclay nanocomposites. Journal of Materials Science, 40 (14) 3783-3787.

John A. Schey,(1987). Introduction to Manufacturing Processes. 2nd Ed., McGraw Hill

Kuilaa T., Bose S., Khanra P., Kim N. H., Rhee K. Y., Lee J. H.(2011).Characterization and properties of in situ emulsion polymerized poly(methyl methacrylate)/grapheme nanocomposites, Composites Part A: Applied Science and Manufacturing, 42 (11) 1856-1861.

Kalaitzidou K., Fukushima H., Drzal L. T. (2010). A new compounding method for exfoliated graphite-polypropylene nanocomposites with enhanced flexural properties and lower percolation threshold. Composites Science and Technology, 67(10) 2045-2051.

Kuilla T., Bhadra S., Yao D., Kim N. H., Bose S., Lee J. H. (2010). Recent advances in graphene based polymer composites. Progress in Polymer Science, 35 (11) 1350-1375.

Koval'chuk A. A., Shchegolikhin A. N., Shevchenko V.G., Nedorezova P. M., Klyamkina A. N., Aladyshev A. M.(2008). Synthesis and properties of polypropylene/multiwall carbon nanotube composites. M acromolecules, , 41 (9) 3149-3156.

Kalaitzidou K., Fukushima H., Drzal L. T.(2007). A new compounding method for exfoliated graphite-polypropylene nanocomposites with enhanced flexural properties and lower percolation threshold. Composites Science and Technology, 67(10) 2045-2051.

Khaled S. M., Sui R., Charpentier P. A., Rizkalla A. S.(2007). Synthesis of TiO_2-PMMA nanocomposite: using methacrylic acid as a coupling agent. Langmuir, 23 (7) 3988-3995.

Kim M., Hong C. K., Choe S., Shim S. E. (2007). Synthesis of polystyrene brush on multiwalled carbon nanotubes treated with $kmno_4$ in the presence of a phase-transfer catalyst. Journal of Polymer Science: Part A: Polymer Chemistry, 45(19) 4413-4420.

Kim B. S., Bae S. H., Park Y.-H., Kim J.-H. (2007). Preparation and Characterization of Polyimide/Carbon-Nanotube Composites. Macromolecular Research, 15(4) 357-362.

Kim B. S., Choi J. S., Lee C. H., Lim S T., Jhon M. S., Choi H. J.(2005) Polymer blend/organoclay nanocomposite with poly(ethylene oxide) and poly(methyl methacrylate). European Polymer Journal, 41(4) 679-685.

Kojima Y., Usuki A., Kawasumi M., Okada A., Kurauchi T., Kamigaito O. (1993). Synthesis of nylon-6-clay hybrid by montmoillinite intercalated with є-caprolactam. Journal of Polymer Science Part A Polymer Chemistry, 31(4)983–986.

Koerner H., Price G., Pearce N. A., Alexander M., Vaia R. A. (2004). Remotely actuated polymer nanocomposites—stress-recovery of carbon-nanotube-filled thermoplastic elastomers. Nature Materials, 3,11-120.

Kojima Y., Usuki A., Kawasumi M., Fukushima Y., Okada A., Kurauchi T., Kamigaito O. (1993). Mechanical properties of nylon 6–clay hybrid. Jouurnal of Material Research, 8, 1179–84.

Li M., Jeong Y. G. (2011). Preparation and Characterization of High performance poly(trimethylene terephthalate)nanocomposites reinforced with exfoliated graphite, Macromolecular Materials and Engineering, 296 (2) 159–167.

Lahelin M., Annala M., Nykänen A., Ruokolainen J., Seppäl J. (2011). In situ polymerized nanocomposites: Polystyrene/CNT and Poly(methyl methacrylate)/CNT composites. Composites Science and Technology, 71 (6) 900-907.

Lentz D., Pyles R., Hedden R.(2010). Surface infusion of gold nanoparticles into processed thermoplastics. Polymer Engineering Science, 50 (1) 120-127.

Lee J.-II, Yang S.B., Jung H.T. (2009). Carbon nanotubes–polypropylene nanocomposites for electrostatic discharge applications, Macromolecules, 42 (21) 8328–8334.

Lee D., Cui. T. (2009). Layer-by-layer self-assembled single-walled carbon nanotubes based ion-sensitive conductometric glucose biosensors. Sensors Journal, IEEE , 9 (4) 449–456.

Luo J J., Daniel. I. M. (2003). Characterization and modeling of mechanical behavior of polymer/clay nanocomposites, Composites Science and Technology, 63(11), 1607–1616.

Li J., Zhou C., Wang G., Zhao D. (2003). Study on kinetics of polymer melt intercalation by a rheological approach, Journal of Applied Polymer Science, 89(2) 318 -323.

Lia J. X., Wu J., Chan C. M. (2000). Thermoplastic nanocomposites. Polymer, 41(18) 6935-6937.

Ma X., Lee N. H., Oh H.J., Hwang J.S., Kim S. J.((2010). Preparation and Characterization of Silica/Polyamide-imide Nanocomposite Thin Films. Nanoscale Research Letters, 5 (111) 846-1851.

Monticelli O., Musina Z., Russo S., Bals S.(2007). On the use of TEM in the characterization of nanocomposites, Materials Letters, 61(16) 3446 -3450.

Mirkin C.A.(2005). The begning of a small revoluation. Small, 1(1) 4-16.

Meincke O., Hoffmann B., Dietrich C., Friedrich C. (2003). Viscoelastic properties of polystyrene nanocomposites based on layered silicates. Macromolecular Chemistry and Physics, 204 (5-6) 823-830.

Nanoscience and Nanotechnologies, July 2004, The Royal Society & the Royal Academy of Engineering.

Omer M., Haider S., Park S. Y. (2011). A novel route for the preparation of thermally sensitive core-shell magnetic nanoparticles. Polymer, 52 (1) 91-97.

Panwar A., Choudhary V., Sharma D.K.(2011). Review: A review: polystyrene/clay nanocomposites. Journal of Reinforced Plastics and Composites, 30(5) 446-459.

Potts J. R., Dreyer D. R., Bielawski C. W., Ruoff R. S. (2011). Graphene-based polymer nanocomposites. Polymer, 52 (1) 5-25.

Patole A. S., Patole S. P., Kang H., Yoo J.-B., Kim T.-H., Ahn J.-H. (2010). A facile approach to the fabrication of graphene/polystyrene nanocomposite by in situ microemulsion polymerization. Journal of Colloid and Interface Science, 350 (2) 530-537.

Paul D R., Robeson L. M. (2008). Polymer nanotechnology: Nanocomposites. Polymer, 49 (15) 3187-3204.

Parka J. H., Lee H M., Chin I.-J., Choi H. J., Kim H. K., Kang W. G. (2008). Intercalated polypropylene/clay nanocomposite and its physical characteristics. Journal of Physics and Chemistry of Solids, 699 (5-6) 1375-1378.

Pattanayak A., Jana S. C. (2005). Synthesis of thermoplastic polyurethane nanocomposites of reactive nanoclay by bulk polymerization methods. Polymer, 46(10) 3275-3288.

Park, C., Park, O., Lim, J., Kim, H. (2001). The fabrication of syndiotactic polystyrene/ organophilic clay nanocomposites and their properties. Polymer, 42 (17) 7465–7475.

Ritzhaupt-Kleissl E., Boehm J., Hausselt J., Hanemann T.(2006) Thermoplastic polymer nanocomposites for applications in optical devices. Materials Science and Engineering: C, 26 (5-7) 1067-1071.

Ray S. S., Okamoto M. (2003). Polymer/layered silicate nanocomposites: a review from preparation to processing. Progress in Polymer Science, 28 (11) 1539-1641.

Silva G. G., Rodrigues M.-T. F., Fantini C., Borges R. S., Pimenta M. A., Carey B. J., Ci L., Ajayan P. M. (2011) Thermoplastic polyurethane nanocomposites produced via impregnation of long carbon nanotube forests. Macromolecular Materials and Engineering, 296 (1) 53-58.

Saeed K., Park S.-Y. (2010). Preparation and characterization of multiwalled carbon nanotubes/polyacrylonitrile nanofibers. Journal of Polymer Research, 17 (4) 535-540.

Saeed K., Park S.Y., Haider S., Baek J. B. (2009). In situ polymerization of multi-walled carbon nanotube/nylon-6 nanocomposites and their electrospun nanofiBers, Nanoscale Research Letters, 4 (1) 39-46.

Strong A. B.(2000). *Plastics, Materials and Processing*. 2nd Ed., Prentice Hall.

Schmidt, D., Shah, D., Giannelis, E.P. (2002). New advances in polymer/layered silicate nanocomposites, Current Opinion in Solid State and Materials Science, 6 (3) 205-212.

Stephen Y. C., Peter R. K., Preston J. Renstrom. (1996). Nanoimprint lithography, Journal of Vacuum Science and Technology B, 14 (6) 4129-4133.

Tsivion D., Schvartzman M., Popovitz-Biro R., von Huth P., Joselevich E. (2011). Guided growth of millimeter-long horizontal nanowires with controlled orientations. Science, 333 (6045) 1003-1007.

Thostenson, E., Li C., and Chou T. (2005). Review nanocomposites in context. Journal of Composites Science & Technology, 65 (3-4) 491–516.

Vaia R. A., Price G., Ruth P. N., Nguyen H. T., Lichtenhan J. (1999). Polymer/layered silicate nanocomposites as high performance ablative materials. Applied Clay Science, 15(1-2) 67-92.

Vaia R. A., Teukolsky R. K., Giannelis E. P.(1994). Interlayer structure and molecular environment of alkylammonium layered silicates. Chemistry of Materials, 6 (7)1017–1022.

Wang F.J., Li W. , Xue M.S., Yao J.P., Lu J.S. (2011). $BaTiO_3$-polyethersulfone nanocomposites with high dielectric constant and excellent thermal stability. Composites Part B: Engineering, 42 (1) 87-91.

Wu. Y., Zhang. Y., Xu. J., Chen. M., Wu. L. (2010). One-step preparation of PS/TiO2 nanocomposite particles via mini emulsion polymerization. Journal of Colloid and Interface Science, 343 (1) 18-24.

XuY., Bing H., Piao J. N., Ye S.-R., Huang J. (2008). Crystallization behavior of Poly(trimethylen terephthalate/multi-walled nanotube composites. Journal of Material Science, 43(1) 417-421.

Xu Z., Gao C. (2010). In situ polymerization approach to graphene-reinforced nylon-6 composites. Macromolecules, 43 (16) 6716–6723.

Xu Y., Brittain W. J., Xue C., Eby R. K.(2004), Effect of clay type on morphology and thermal stability of PMMA–clay nanocomposites prepared by heterocoagulation method. Polymer, 45(11) 3735–3746.

Yu S., Juay Y.K., Young M. S.(2008). Fabrication and characterization of carbon nanotube reinforced poly(methyl methacrylate) nanocomposites. Journal Nanoscience and Nanotechnology, 8 (4) 1852-1857.

Zhao W., Li M., Peng H.X. (2010). Functionalized MWNT-Doped Thermoplastic Polyurethane Nanocomposites for Aerospace Coating Applications. Macromolecular Materials and Engineering, 295(9) 838-845.

7

Crystallization and Thermal Properties of Biofiber-Polypropylene Composites

M. Soleimani[1], L. Tabil[1], S. Panigrahi[1] and I. Oguocha[2]
[1]Department of Chemical and Biological Engineering, University of Saskatchewan
[2]Department of Mechanical Engineering, University of Saskatchewan
Saskatoon,
Canada

1. Introduction

The crystallization of small molecules proceeds by nucleation and growth mechanisms. In polymers, the basic morphology of the crystals is spherulite lamellar crystal bundles which results from the growth of a nucleus center followed by branching to form radial structural equivalence. According to reported observations (Reiter & Strobl 2007), the long-chain fractions in spherulites enriched in early-formed thick crystals are called dominant lamellae and the short-chain fractions enriched in later-formed thin crystals are called subsidiary lamellae.

Polypropylene (PP) is an attractive thermoplastic polymer with exceptional properties such as high isotacticity, high cost-performance ratio, low processing temperature, excellent chemical and moisture resistance, low density and good ductility (Somnuk et al. 2007 ; Zhang et al. 2002). However, it has some inferior mechanical properties such as low impact resistance and low stiffness, both of which can be improved upon by using additives such as tougheners and the application of nucleating agents (Zhang et al. 2002).

The application of nucleating agents results in the shortening of injection molding cycle and, consequently, in the reduction of manufacturing costs. Also, optical and mechanical properties of polymers can be improved by the generation of small spherulites. As a common industrial practice, polymers are often mixed with other materials to improve properties such as strength and biodegradability or to save the starting materials (Mucha & Krolikowski 2003).

As a coupling agent used for in situ or reactive compatibilization, maleic anhydride-grafted polypropylene (MAPP) has the same molecular structure as polypropylene while the maleic anhydride group is attached to the backbone. In a study by Seo et al. (2000), it was reported that the mechanism of crystallization in isotactic polypropylene (iPP) could be different from that of MAPP due to their different nucleation mechanisms originating from the differences in their characteristics and the number of heterogeneous nuclei. Also, the diffusional activation energy and crystallization half-time were found to be smaller for MAPP than for iPP under isothermal conditions. The rate of crystallization was decreased by increasing temperature under isothermal conditions for both materials, however, it was much more noticeable for iPP. The application of MAPP in iPP affected its crystallization during the cooling process because of the increase in the number of effective nuclei.

In this paper, non-isothermal crystallization melt behaviour and thermal properties of PP composite materials with different formulations were studied with respect to the effects of chemical modification, the use of compatibilizer, and fiber loading. For this purpose, different theories and models were used to analyze the data obtained in this investigation.

2. Materials and methods

2.1 Materials

A compression-grade PP (PRO-FAX) with density of 0.904 g cm^{-3}, melt flow index of 0.65 g/10 min at 230°C, and low melt flow index (MFI = 0.65 g/10 min at 230°C) was obtained from Ashland Specialty Chemical Company (Vancouver, BC, Canada), while maleic anhydride-grafted polypropylene (MAPP) (MFI = 115 g/10 min at 190°C; maleic acid content of approximately 0.6%) was obtained from Aldrich Chemical Company (Toronto, ON, Canada). Flax fiber which was already retted with a density of 1.52 g cm^{-3} was obtained from Biofiber Ltd., Canora, SK, Canada. For mercerization, the fiber was first washed with a 2% commercially available detergent solution (Ultra liquid Tide containing cationic and non-ionic biodegradable detergents) and then washed with distilled water to eliminate extractives, especially waxy materials. After drying at 60°C for 24 h, it was pretreated with a 5% sodium hydroxide (NaOH) solution for 3 h and thoroughly washed with distilled water and dried again in an oven (Despatch Oven Co., Minneapolis, MN, USA). Fiber analysis for measurement of the components was performed based on the measurement of neutral detergent fiber (NDF), acid detergent fiber (ADF) and acid detergent lignin (ADL) using a fiber analyzer (ANKOM Technology, Fairport, NY, USA) to determine cellulose, hemicellulose, lignin and ash percentages.

2.2 Fabrication of composites

Pretreated and untreated flax fibers were each milled in a grinder (Retsch GmbH 5657 HAAN, West Germany) through a 2-mm opening to be used in composite formulations as shown in Table 1. PP and MAPP were dried in the oven at 60°C for 15 h and at 120°C for 15 h, respectively, before use. Materials based on the formulation and after initial mixing were extruded in a single-screw extruder (Akron Inc., Batavia, OH, USA) at temperatures up to 190°C with a screw speed of 45 rpm and the extrudates were pelletized to be used for compression molding in a hot press (J.B. Miller Machinery & Supply Co., Toronto, ON, Canada) under a pressure of 3.5 MPa at 190°C for 7 min to prepare plates with a thickness of about 3.2 mm. Differential scanning calorimetry (DSC) measurements were carried out from small pieces cut from moulded composites.

2.3 Thermal conductivity and density measurements

A thermal conductivity analyzer (FOX 200, Saugus, MA, USA) was used to determine the steady-state effective thermal conductivity of the molded polymer and composites at 25°C in accordance with ASTM C518. 2002. Each run took 0.5 h, but the first 5 min was used to bring the samples to the steady-state condition. The density of the test materials was measured by using a gas (nitrogen)-operated pycnometer (Quantachrome Corp., Boynton Beach, FL, USA) to measure the volume of the samples and their mass was determined using a Galaxy 160D weighing scale (OHAUS Scale Corporation, Florham Park, NJ, USA).

2.4 Differential scanning calorimetry

DSC (TA Instruments, New Castle, DE, USA) measurements were performed in a TA Instrument model 2000 DSC equipped with a cooling system to assess crystallization properties of the materials. Samples were heated from 40°C to 200°C at a heating rate of 10°C/min and held for 5 min to erase the thermal history of the polymer. Then, the samples were cooled down at the desired rate (5, 10, 15 and 20°C/min) to analyze and investigate the crystallization kinetics. The degree of crystallinity (X_c) in biocomposites corrected for biofibers was determined by integration of the generated DSC exotherms. The crystallinity of PP or the matrix in the composites was calculated using equation (1).

$$X_c\% = \frac{\Delta H_c}{\Delta H_c^0} \times 100 \quad (1)$$

Where ΔH^o_c is the heat of crystallinity of 100% crystalline PP assumed to be 146.5 J/g (Lonkar, et al. 2009) and ΔH_c is taken as the enthalpy of crystallization corrected for biofiber in the composites assuming that the contribution of this fraction is ignored.

Fiber	PP/MAPP/Fiber	Formulation (%)
-	PP	100/0/0
Untreated fiber	C1	85/0/15
	C2	80/5/15
	C3	70/0/30
	C4	65/5/30
Alkaline treated fiber	C5	85/0/15
	C6	80/5/15
	C7	70/0/30
	C8	65/5/30

Table 1. Components of the composites based on polypropylene and flax fiber.

3. Results and discussion

Chemical analysis of the (retted) flax fiber before pretreatment showed that the mass fractions of cellulose, hemicellulose and lignin were 80.9%, 7.9% and 1.4%, respectively. These changed to 85%, 6.2% and 1.2%, respectively, after mercerization.

3.1 Thermal conductivity

The thermal conductivity values determined at 25°C for the slab-shaped test materials are given in Table 2. It can be seen that thermal conductivity decreased in all composites compared to the unreinforced plain PP which means that flax fiber can increase the thermal insulation property of polymers. The reduction in thermal conductivity is due to the inherent low thermal conductivity of cellulosic materials in comparison to the unreinforced plain polymer. A close observation shows that the thermal conductivity of the composites C6 and C8 (those containing treated fiber, plain PP and MAPP) was slightly higher than similar composites reinforced with untreated fiber (i.e., C2 and C4). This is probably due to the ability of MAPP to improve cross links between the fiber and matrix as reported by Kim et al. (2006).

Fiber	Material	Density ($g\ cm^{-3}$)	Thermal conductivity (W/m°C)
-	PP	0.901	0.152
Untreated fiber	C1	0.956	0.133
	C2	0.957	0.133
	C3	1.006	0.126
	C4	1.013	0.126
Alkaline treated fiber	C5	0.964	0.135
	C6	0.963	0.137
	C7	1.021	0.130
	C8	1.022	0.133

Table 2. Density and thermal conductivity of PP (polypropylene) and composites(C1 to C8 as indicated by Table 1).

3.2 Crystallization behavior

Figure 1 shows the DSC exotherms obtained for samples cooled from the melt at different cooling rates (5 to 20°C/min). The crystallization temperature (T_c) and crystallinity (X_c%) of the test materials are presented in Table 3.

The effect of cooling rate on the shape and relative position of the peak temperature (i.e., crystallization temperature, T_c) of the exotherms can be readily discerned from Figure 1 and Table 3. All the curves shifted to lower temperatures as heating rate increased. It can be observed that the higher the cooling rate, the lower the crystallization temperature and degree of crystallinity. The data indicate that the average shift in crystallization temperature for composites is approximately 6°C as the cooling rate increased from 5 to 20°C/min. For pure PP, it is about 4°C. The crystallinity of both pure PP and composites decreased with the cooling rate apparently because the low cooling rates provide higher fluidity and diffusivity for the polymer matrix molecules, thereby improving secondary crystallization and inducing more crystallinity at high temperatures than at high cooling rates. Furthermore, the results also indicate that the addition of fiber increased crystallinity of the test materials. This is attributed to the nucleation effect of the fibers which provide nucleation sites and facilitate crystallization of the polymer as well as transcrystallinity (Somnuk et al. 2007). It can also be discerned from Table 3 that the contribution of biofiber in the composites not only increased crystallinity of the material, but also increased crystallization temperature at the same cooling rate (Table 3).

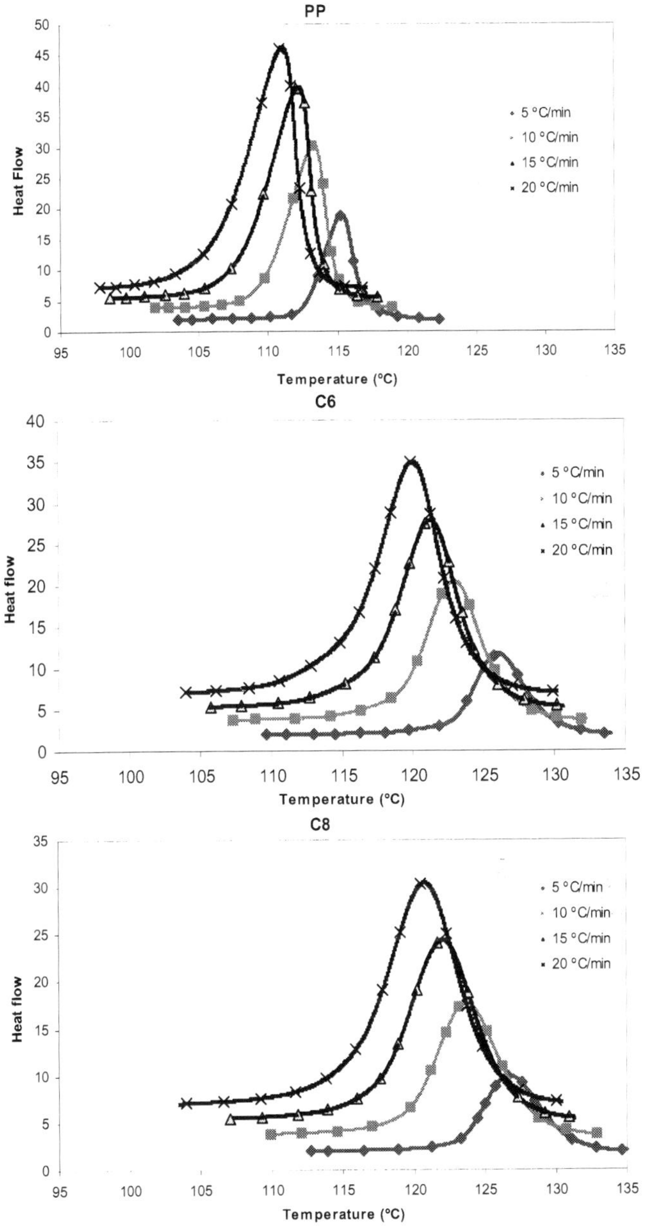

Fig. 1. Non-isothermal crystallization thermograms of PP (polypropylene), C6 (PP/MAPP/Fiber: 80/5/15) and C8 (PP/MAPP/Fiber: 65/5/30) at different cooling rates.

Formulation	5 °C/min		10 °C/min		15 °C/min		20 °C/min	
	T_c (°C)	X_c(%)	T_c (°C)	X_c(%)	T_c (°C)	X_c(%)	T_c (°C)	X_c(%)
PP	115.3	59.5	113.4	59.1	112.2	59.6	111.1	58.2
C1	123.4	63.1	120.1	62.9	118.3	62.1	116.9	61.2
C2	125.1	62.3	122.3	62.5	120.5	61.0	119.3	61.0
C3	124.8	66.4	121.7	65.9	120.0	65.2	118.7	65.1
C4	126.1	63.2	123.1	62.0	121.4	61.0	120.1	60.0
C5	122.6	62.1	119.2	61.9	117.3	60.5	115.9	60.2
C6	125.3	61.4	122.0	60.3	120.4	60.1	119.1	60.0
C7	124.5	65.2	121.0	64.2	119.3	64.0	118.1	63.1
C8	126.3	62.8	123.3	61.8	121.5	60.9	120.3	60.5

PP: polypropylene; C_1 to C_8: composites; T_c: crystallization temperature; X_c: crystallinity.

Table 3. Crystallization temperature and crystallinity of PP and biocomposites at different cooling rates (5, 10, 15 and 20 °C/min).

The use of the compatibilizer seems to have reduced the level of crystallinity in the composites for the same levels of fiber loading, while it resulted in a marginal increase in the crystallization temperature. For instance, at the cooling rate of 5°C/min, the crystallization temperature increased from 124.5°C to 126.3°C, while the crystallinity decreased from 65.2% to 62.8% for C7 and C8, respectively; and this could be attributed to the application of MAPP in C8. In the same vein, chemical pretreatment seems to marginally reduce the degree of crystallinity for all composites irrespective of their fiber content in comparison with composites reinforced with the untreated fiber with or without the compatibilizer. Therefore, among all formulations, the highest value of crystallinity was obtained for composite sample C3 at the cooling rate of 5°C/min.

Fiber mercerization also slightly caused the reduction of crystallization temperature at the constant level of fiber loading as well as constant cooling rate only for samples without MAPP, but it almost did not have any influence on crystallization temperature with the presence of MAPP. For example, at the cooling rate of 10°C/min, crystallization temperature changed from 120.1°C for C1 to 119.2°C for C5 and this is only because of the chemical modification.

The values of onset temperature (T_0), end temperature of crystallization exotherm (T_e), peak time (t_c) and half crystallization time ($t_{0.5}$) determined for all samples at different cooling rates are summarized in Table 4. It can be seen that these four parameters decreased with increasing cooling rate. The application of MAPP in the composites reduced the half-time of crystallization for composites reinforced with pretreated and untreated biofiber. For example, $t_{0.5}$ decreased from 55 to 48 s for C7 and C8 at 5°C/min, respectively. However, the magnitude of reduction in half-time due to the compatibilizer was less at high cooling rates. Also, a comparison of the data in Table 4 indicates that chemical pretreatment of the fiber increased the magnitude of $t_{0.5}$ in the biocomposites at the same level of fiber content and cooling rate. This result is consistent with those reported by Garbarczyk et al. (2000) who observed that PP crystallized faster when reinforced with untreated natural fiber than with chemically modified fibers. Furthermore, a close look at the data shows that lowest value of half crystallization time occurred in pure PP for all cooling rates. Although its crystallinity is lower at all cooling rates, it is observed that non-isothermal crystallization occurred faster in PP than in the biocomposites.

Formulation	5 °C/min				10 °C/min				15 °C/min				20 °C/min			
	T_0 (°C)	T_e (°C)	t_c (s)	$t_{0.5}$ (s)	T_0 (°C)	T_e (°C)	t_c (s)	$t_{0.5}$ (s)	T_0 (°C)	T_e (°C)	t_c (s)	$t_{0.5}$ (s)	T_0 (°C)	T_e (°C)	t_c (s)	$t_{0.5}$ (s)
PP	116.9	103.3	29	26	114.9	101.9	17	16	113.7	98.5	13	13	112.7	97.8	10	10
C1	128.0	111.4	59	55	125.2	104.2	34	31	123.5	103.3	25	23	122.3	102.1	19	18
C2	128.8	112.5	50	46	125.9	106.1	26	25	124.2	106.1	19	18	122.9	104.7	14	14
C3	128.6	112.8	50	49	125.9	107.8	29	28	124.3	105.9	21	20	123.2	103.1	17	16
C4	130.1	109.7	53	48	127.3	110.2	29	26	125.7	107.8	20	19	124.4	104.5	16	15
C5	127.4	110.2	61	57	124.4	104.9	34	32	122.7	104.7	25	23	121.5	102.1	20	19
C6	129.2	111.8	52	48	126.1	108.3	29	27	124.4	106.4	20	19	123.1	103.8	16	15
C7	128.9	111.1	57	55	125.9	109.5	32	31	124.3	105.9	23	22	123.2	103.5	18	17
C8	130.4	112.8	53	48	127.6	123.3	30	28	126.0	109.0	21	19	124.8	105.2	17	16

PP: polypropylene; C_1 to C_8: composites; T_0: onset temperature, T_e: end temperature of crystallization, t_c: peak time and $t_{0.5}$: half crystallization time.

Table 4. Crystallization parameters obtained from DSC exotherms.

3.3 Modeling of crystallization

The relative degree of crystallinity (X_t) was calculated from the relationship in equation (2) as:

$$X_t = \frac{\int_{T_0}^{T} (dH_c/dT)dT}{\int_{T_0}^{T_\infty} dH_c / dT)dT} \tag{2}$$

where dH denotes the enthalpy of crystallization measured during the time interval dt and T_∞ is the end temperature of crystallization. Figure 2 shows typical plots of relative crystallinty vs. temperature obtained for representative samples of the tested materials (PP, C6 and C8). Similar plots were obtained for all samples tested in this study. It can be seen that the shape of the curves is sigmoidal. It can be observed from these plots that, for the cooling rates studied, PP crystallized at lower temperatures than the composite materials.

The activation energy for crystallization (ΔE) was determined for pure PP and the biocomposite samples using the Kissinger model (1957) shown in equation (3).

$$\frac{-\Delta E}{R} = \frac{d(\ln(\varphi/T_c^2))}{d(1/T_c)} \tag{3}$$

where R is the universal gas constant (kJ/mol.K) and φ is the cooling rate. The values of activation energy for PP and its composites were determined from the slope of the linear plots shown in Figures 3 and 4. It can be seen from Table 4 that the addition of biofiber markedly reduced the activation energy for crystallization in pure PP. The low activation energy for crystallization obtained for the composites is consistent with their high crystallinity reported in Table 3, which could be attributed to the fact that biofibers, being high energy sites, lowered the activation energy for nucleation as similarly reported by Page & Gopakumar (2006). A close inspection of the data in Table 4 for the composites shows that at each level of fiber content, the application of MAPP increased their activation energy. For instance, it was increased from 287 to 306 kJ/mol for C7 and C8, respectively.

The Avrami model given in equation (4) is used extensively to evaluate the isothermal kinetics of polymer crystallization:

$$1 - X_t = \exp\left(-kt^n\right) \tag{4}$$

where X_t is the relative crystallinity at time t and constant temperature; k is the crystallization rate constant containing the nucleation and growth rates and is temperature dependent; and n is the Avrami index or exponent which depends on the type of nucleation and growth process. Ozawa (1971) extended the Avrami model to non-isothermal processes by assuming that they are the result of infinitely small changes in the isothermal crystallization steps and obtained:

$$1 - X_t = \exp\left(-x / \varphi^m\right) \tag{5}$$

where m is the Ozawa exponent which is dependent on the nucleation density and the spherulitic radial growth rate and x is a function of the overall crystallization rate.

The linearized form of equation (5) is given as:

$$\ln\left[-\ln\left(1 - X_t\right)\right] = \ln x - m \ln\varphi \tag{6}$$

Hence, a plot of $\ln[-\ln(1 - X_t)]$ vs. $\ln(t)$ has a slope of m and a y-intercept of $\ln(x)$.

By combining the Avrami and Ozawa models, Liu et al. (1997) introduced another crystallization kinetic model:

$$\ln\varphi = \ln F(T) - a \ln t \tag{7}$$

where $F(T) = [x/k]^{1/m}$ refers to the crystallization kinetic parameter and a is the ratio of the exponents in Avrami and Ozawa models : $a = n/m$

Typical results of kinetic analysis of the DSC data using the Ozawa model are shown in Figure 5 for PP, C6 and C8, while Table 5 summarizes the results for all formulations.

As shown, the maximum value of x for pure PP is much higher than the values obtained for composites indicating that PP crystallized faster than the matrix in the composites. Amongst the composites, it can be observed in Table 5 that the value of x was affected by fiber content or by compatibilizer for a given fiber content. The parameter m decreased with decreasing temperature for all test materials. Its value ranged from 0.6 to 5.0 for PP and from 1 to 4.2 for the composites. The more limited range of m values obtained for the composites indicates that the crystal growth rate of spherulites was higher in PP than in the composites. This agrees with the result reported by Somnuk et al. (2007) for PP and natural fiber-based composites such that the spherulitic growth rate was higher in neat PP than composites. However, in their study, composites exhibited a higher rate of crystallization compared to neat PP which is different from the result in this study. The F(T) values obtained from the Liu et al. (1997) model increased systematically with the relative crystallinity of pure PP and composites as shown in Table 6. Also, at a given relative crystallinity, the values of F(T) are lower for PP than for the composites at most levels of X_t which indicates that crystallization was faster in pure pp compared to the composites. Furthermore, the application of MAPP in the composites resulted in low values of F(T) or higher crystallization rates.

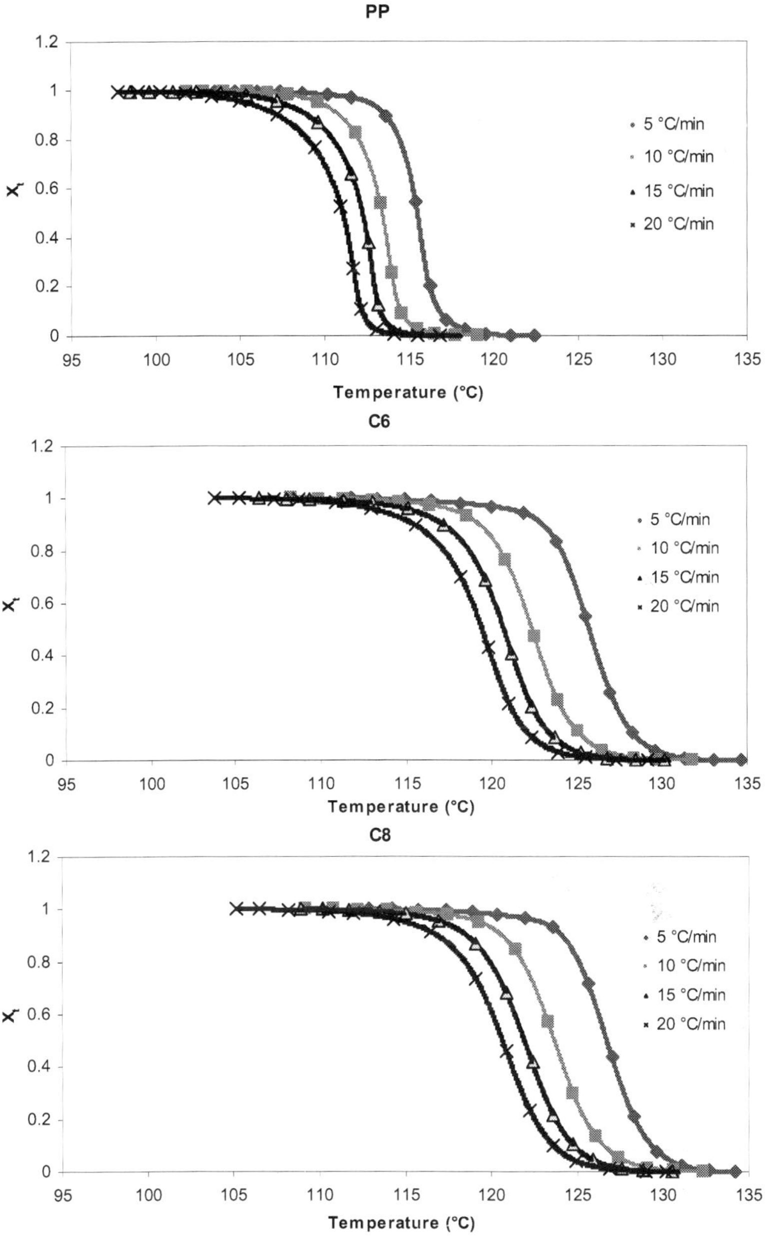

Fig. 2. Relative crystallinity of PP (polypropylene), C6 (PP/MAPP/Fiber: 80/5/15) and C8 (PP/MAPP/Fiber: 65/5/30) at different cooling rates.

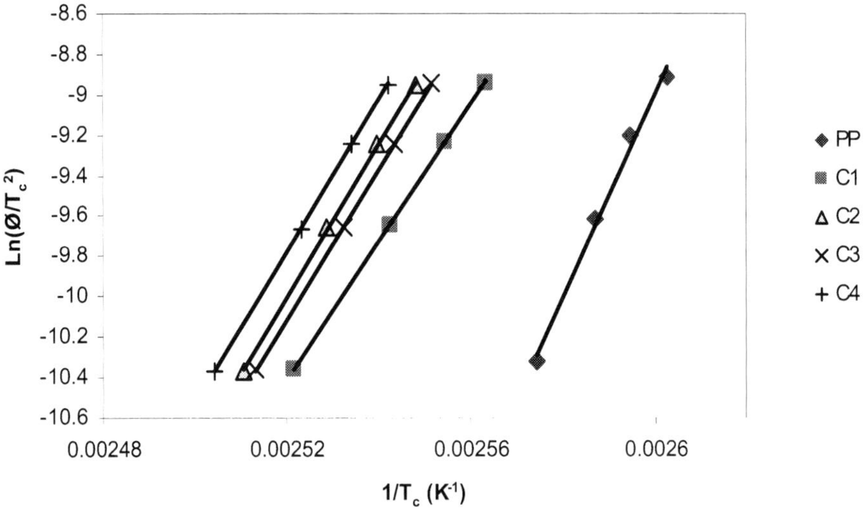

Fig. 3. Kissinger plots of crystallization activation energies of PP (419 kJ/mol) and non-compatibilized formulations: C1 (292 kJ/mol), C2 (314 kJ/mol), C3 (297 kJ/mol) and C4 (310 kJ/mol). PP: polypropylene ; C1, C2, C3 and C4: composites formulations.

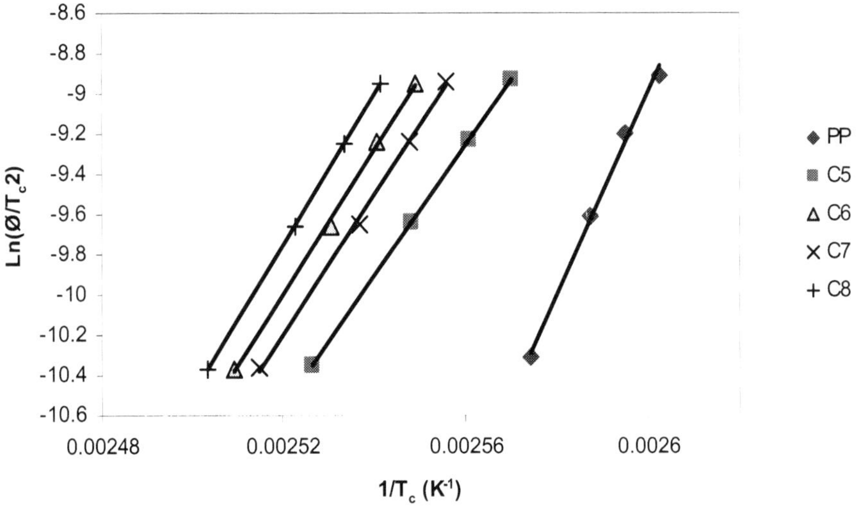

Fig. 4. Kissinger plots of crystallization activation energies of PP (419 kJ/mol) and compatibilized formulations: C5 (272 kJ/mol), C6 (297 kJ/mol), C7 (287 kJ/mol) and C8 (306 kJ/mol). PP: polypropylene ; C5, C6, C7 and C8: composites formulations.

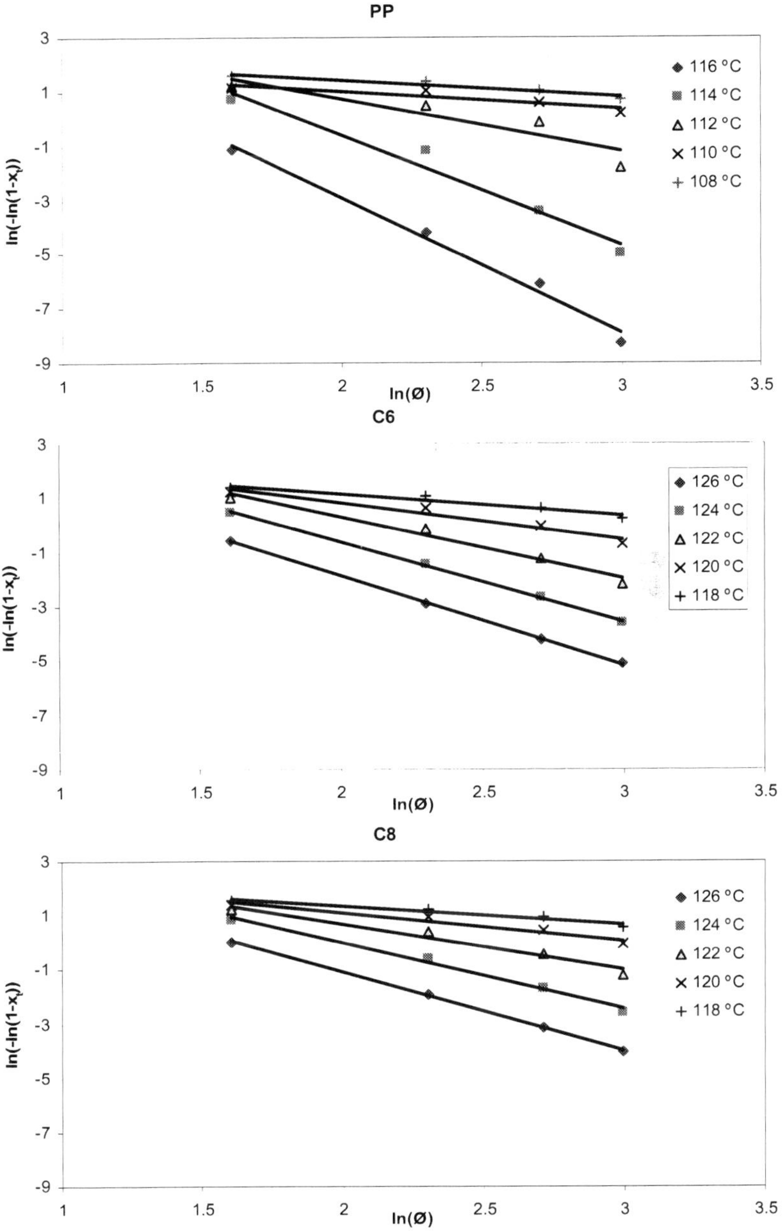

Fig. 5. Ozawa plots for non-isothermal melt crystallization of polypropylene (PP) and composites : C6 (PP/MAPP/Fiber: 80/5/15) and C8 (PP/MAPP/Fiber: 65/5/30).

Material	T(°C)	m	x	R^2
PP	108	0.6	14.9	0.92
	110	0.7	10.0	0.83
	112	1.9	99.5	0.84
	114	4.1	1998.2	0.96
	116	5.0	1212.0	0.99
C1	118	1.4	40.4	0.97
	120	2.0	81.5	0.99
	122	2.5	99.5	0.99
	124	2.9	73.4	1.00
	126	3.4	49.4	1.00
C2	118	0.8	16.4	0.96
	120	1.3	33.1	0.94
	122	2.2	109.9	0.96
	124	2.9	164.0	1.00
	126	3.3	99.5	1.00
C3	118	1.0	22.2	0.98
	120	1.5	40.4	0.97
	122	2.3	99.5	0.98
	124	3.0	148.4	1.00
	126	3.6	121.5	0.99
C4	118	0.7	14.9	0.92
	120	1.0	22.2	0.93
	122	1.8	66.7	0.96
	124	2.5	148.4	0.99
	126	3.1	134.3	1.00
C5	118	1.8	66.7	0.99
	120	2.3	109.9	0.99
	122	2.8	109.9	1.00
	124	3.3	81.5	1.00
	126	4.2	109.9	0.99
C6	118	0.8	16.4	0.94
	120	1.3	36.6	0.95
	122	2.3	121.5	0.98
	124	2.9	200.3	1.00
	126	3.3	109.9	1.00
C7	118	1.1	27.1	0.97
	120	1.7	44.7	0.98
	122	2.3	90.0	0.99
	124	2.7	81.5	1.00
	126	3.0	36.6	1.00
C8	118	0.7	14.9	0.93
	120	1.0	22.2	0.94
	122	1.7	60.3	0.96
	124	2.4	134.3	0.99
	126	2.9	109.9	1.00

PP: polypropylene; C1 to C8: composites; m and x: Ozawa constants, and R^2: coefficient of determination.

Table 5. Kinetic parameters obtained for PP and the composites, T: temperature; using the Ozawa model.

Material	x_t(%)	F(T)	B	R^2
PP	20	6.7	1.0	0.99
	40	7.4	1.0	0.99
	60	8.2	1.0	0.99
	80	9.0	1.1	0.99
C1	20	7.4	1.2	1.00
	40	10.0	1.2	1.00
	60	11.0	1.2	1.00
	80	13.5	1.2	1.00
C2	20	6.7	1.2	1.00
	40	9.0	1.2	1.00
	60	10.0	1.2	1.00
	80	11.0	1.1	1.00
C3	20	6.1	1.3	0.99
	40	8.2	1.3	0.99
	60	9.0	1.3	1.00
	80	11.0	1.3	1.00
C4	20	6.7	1.1	0.99
	40	9.0	1.1	1.00
	60	10.0	1.1	1.00
	80	11.0	1.1	1.00
C5	20	8.2	1.1	0.95
	40	10.0	1.2	0.97
	60	11.0	1.2	0.97
	80	13.5	1.2	0.98
C6	20	8.2	1.1	1.00
	40	10.0	1.1	1.00
	60	11.0	1.1	1.00
	80	12.2	1.1	1.00
C7	20	7.4	1.3	1.00
	40	10.0	1.3	1.00
	60	11.0	1.3	1.00
	80	13.5	1.3	1.00
C8	20	6.7	1.3	0.99
	40	8.2	1.3	1.00
	60	10.0	1.3	1.00
	80	12.2	1.3	1.00

Table 6. Kinetic parameters obtained for PP (polypropylene) and its composites (C1 to C8) using the model of Liu et al. (1997).

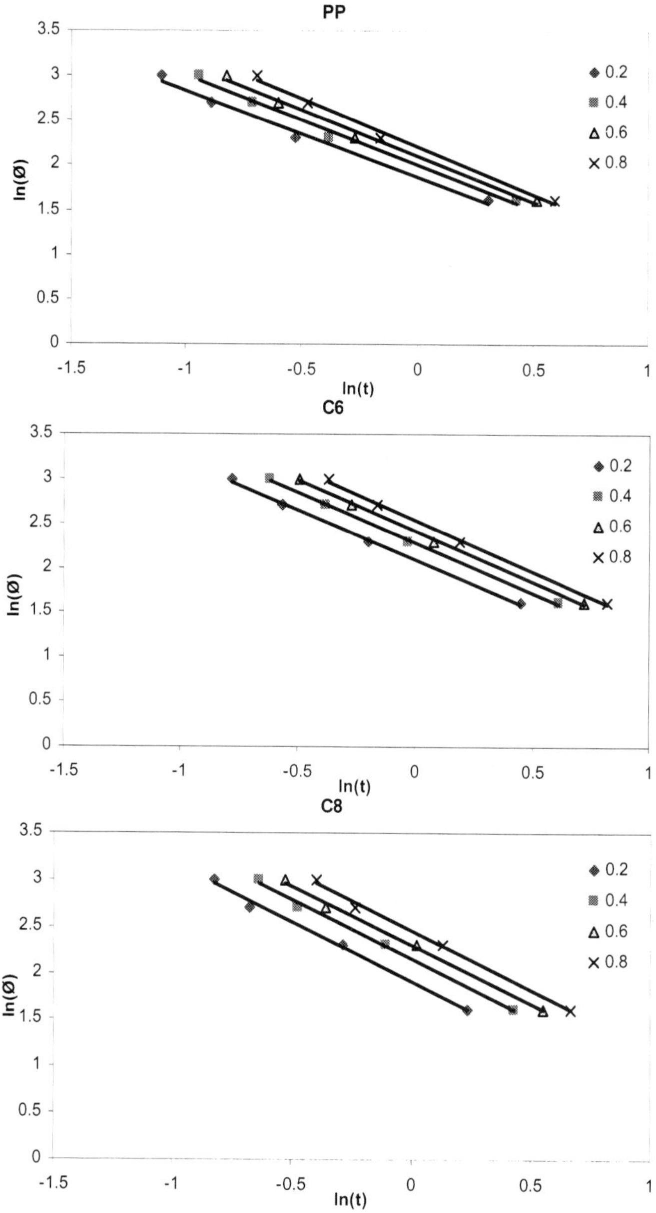

Fig. 6. Non-isothermal crystallization plots based on Liu et al. (1997) model. PP: polypropylene; C6 (PP/MAPP/Fiber: 80/5/15) and C8: (PP/MAPP/Fiber: 65/5/30). composites.

4. Conclusions

From the experimental results, the following conclusions can be drawn:

1. The thermal conductivity and non-isothermal crystallization kinetics of biofiber-reinforced PP composites were influenced by fiber content and chemical modifications of the biofiber.
2. The addition of fiber reduced the thermal conductivity of pure PP.
3. The biocomposites exhibited higher crystallinity, crystallization temperature, half-time but lower crystallization rate than pure PP,.
4. Composites fabricated with chemically-modified fibers exhibited lower degree of crystallinity than those reinforced with untreated fibers.
5. The addition of MAPP into the composites accelerated the crystallization process but had a negative impact on the degree of crystallinity.
6. The important process parameter of cooling rate also was strongly effective on the behaviour of crystallization in that involving higher cooling rate resulted to lower degree of crystallinity and lower crystallization temperature, but accelerated crystallization process. Activation energy of crystallization determined using Kissinger model for composites (around 300 kJ/mol) was much lower than that of PP (419 kJ/mol) in this study which is consistent with the degrees of crystallinity.
7. Analyzing the data using Ozawa and Liu et al. models resulted to a good linearity and conclusion.

5. References

ASTM C518. 2002. Standard Test Method for Steady-State Heat Flux Measurements and Thermal Transmission Properties by Means of the Heat Flow Meter Apparatus, Philadelphia, PA.

ANKOM Technology (2005) ANKOM Technology Method. http://www.ankom.com/09_procedures/ procedures.shtml .

Garbarczyk, J. & Borysiak, S. (2000). Crystallization of isotactic polypropylene at surfaces of cellulose natural fibers. *3rd International Wood and Natural Fibre Composites Symposium* Kassel, Germany.

Kim, S.W.; Lee, S.H.; Kang, J.S. & Kang, K.H. (2006). Thermal conductivity of thermoplastics reinforced with natural fibers. *International Journal of Thermophysics* vol.27, No.6, pp. 1873-1881, ISSN 1572-9567.

Kissinger, H.T. (1957). Reaction Kinetics in Differential Thermal Analysis. *Analytical Chemistry* Vol.29, No.11, pp. 1702-1706, ISSN 0003-2700.

Liu, T.; Mo, Z.; Wang, S. & Zhang, H. (1997). Nonisothermal melt and cold crystallization kinetics of poly(aryl ether ether ketone ketone). *Polymer Engineering & Science* Vo.37, No.3, pp. 568-575, ISSN 1548-2634.

Lonkar, S.P.; Morlat-Therias, S.; Caperaa, N.; Leroux, F,; Gardette, J.L. & Singh, R.P. (2009). Preparation and nonisothermal crystallization behavior of polypropylene/layered double hydroxide nanocomposites. Polymer Vol.50, No.6, pp. 1505-1515, ISSN 0032-3861.

Mucha, M. & Krolikowski. (2003). Application of DSC to study crystallization kinetics of polypropylene containing fillers. *Journal of Thermal Analysis and Calorimetry* Vol.74, pp. 549-557, ISSN 1572-8943.

Ozawa, T. (1971). Kinetics of non-isothermal crystallization, *Polymer* 12(3): 150-158.

Page, D.J.Y.S. & T.G. Gopakumar. (2006). Properties and crystallization of maleated polypropylene/graphite flake nanocomposites. *Polymer Journal* Vol.38, No.9, pp. 920-929, ISSN 0032-3896.

Reiter, G. & Strobl, G.R. (2007). *Progress in understanding of polymer crystallization*. Springer-Verlag Berlin Heidelberg. NY, USA.

Seo, Y.; Kim, J.; Kim, K.U. & Kim, Y.C. (2000). Study of the crystallization behaviors of polypropylene and maleic anhydride grafted polypropylene. *Polymer* Vol.41, pp. 2639-2646, ISSN 0032-3861.

Somnuk, U.; Eder, G.; Phinyocheep, P.; Suppakarn, N.; Sutapun, W. & Ruksakulpiwat, Y. (2007). Quiescent crystallization of natural fibers-polypropylene composites. *Journal of Applied Polymer Science* Vol.106, pp. 2997–3006, ISSN 1097-4628.

Zhang, X.; Xie, F.; Pen, Z. & Zhang, Y. & Zhou W. (2002). Effect of nucleating agent on the structure and properties of polypropylene/poly (ethylene–octene) blends. *European Polymer Journal* Vol.38, No.1, pp. 1-6, ISSN 0014-3057.